T0313014

# Breakthroughs in Smart City Implementation

# RIVER PUBLISHERS SERIES IN COMMUNICATIONS

*Series Editors*

**ABBAS JAMALIPOUR**
*The University of Sydney*
*Australia*

**JUNSHAN ZHANG**
*Arizona State University*
*USA*

**MARINA RUGGIERI**
*University of Rome Tor Vergata*
*Italy*

Indexing: All books published in this series are submitted to Thomson Reuters Book Citation Index (BkCI), CrossRef and to Google Scholar.

The "River Publishers Series in Communications" is a series of comprehensive academic and professional books which focus on communication and network systems. The series focuses on topics ranging from the theory and use of systems involving all terminals, computers, and information processors; wired and wireless networks; and network layouts, protocols, architectures, and implementations. Furthermore, developments toward new market demands in systems, products, and technologies such as personal communications services, multimedia systems, enterprise networks, and optical communications systems are also covered.

Books published in the series include research monographs, edited volumes, handbooks and textbooks. The books provide professionals, researchers, educators, and advanced students in the field with an invaluable insight into the latest research and developments.

Topics covered in the series include, but are by no means restricted to the following:

- Wireless Communications
- Networks
- Security
- Antennas & Propagation
- Microwaves
- Software Defined Radio

For a list of other books in this series, visit www.riverpublishers.com

# Breakthroughs in Smart City Implementation

## Editors

### Leo P. Ligthart

CONASENSE, The Netherlands

### Ramjee Prasad

GISFI, India
CTIF Global Capsule, Aarhus University, Denmark

Routledge
Taylor & Francis Group

LONDON AND NEW YORK

**Published 2017 by River Publishers**
River Publishers
Alsbjergvej 10, 9260 Gistrup, Denmark
www.riverpublishers.com

**Distributed exclusively by Routledge**

4 Park Square, Milton Park, Abingdon, Oxon OX14 4RN
605 Third Avenue, New York, NY 10158

First published in paperback 2024

*Breakthroughs in Smart City Implementation* / by Leo P. Ligthart, Ramjee Prasad.

*Routledge is an imprint of the Taylor & Francis Group, an informa business*

Publisher's Note
The publisher has gone to great lengths to ensure the quality of this reprint but points out that some imperfections in the original copies may be apparent.

While every effort is made to provide dependable information, the publisher, authors, and editors cannot be held responsible for any errors or omissions.

ISBN: 978-87-999237-2-4 (hbk)
ISBN: 978-87-7004-433-2 (pbk)
ISBN: 978-1-003-33740-9 (ebk)

DOI: 10.1201/9781003337409

# Contents

## 7   Challenges in the Design of Smart Vehicular Cyber Physical Systems with Human in the Loop    165

*Agata Manolova, Vladimir Poulkov and Krasimir Tonchev*

## 8   What Makes Cities Bloom and Prosper?: Connected & Cooperating People    189

*Jaap van Till*

## 9   Smart Cities – A Panacea for the Ills of Urbanization: An Indian Perspective    215

*M. D. Lele*

# Foreword

Foundation on Communication, Navigation, Sensing, and Services (hereinafter "CONASENSE") is a scientific society with the vision on Communication, Navigation, Sensing, and Services (CNSS), 20 to 50 years from now. A multitude of future services for our information society can get benefit from the integration of Communications, navigation and sensing systems; some examples are in the area of: Air, Road, and Maritime Traffic Management, real-time alert systems, Earth science, and interplanetary space science, disaster monitoring, safety critical services, and many more.

Sensors play a key role in several communications and navigation applications. An example is the wide variety of sensor applications in road-, rail-, maritime-, air- and space-based navigation used in traffic systems. Other sensor applications important in Conasense are medical sensors in health monitoring, photonic sensors for robots in the area of universe observations, inertial optoelectronic sensors for Earth observation, photonic nanosensors for scientific payloads, and optoelectronic sensors for stress and vibration detection in transport systems, inertial nanoelectronic sensors for satellite navigation, etc. Several of these sensor applications have got attention in the previous four Conasense books.

This 5th book has its focus on Smart City. In the vision of Conasense members this broad topic is connected to many social, political, and technological issues which need urgent solutions. At the same time, it is noted that full solutions will not occur within 5 to 10 years. Therefore, it is good that Conasense does "brain storming" on areas as green and long lasting solutions for improving air quality, for quality of life of residents in cities and solutions for traffic congestions. Conasense has established two branches, one in China and one in India. These 2 countries have understood that Smart City programs are essential in order to overcome the obvious shortcomings in the moving of millions of people to cities. As chairman and Founder member of CONASENSE the significant contribution to this book by both branches with 3 Chapters from China and 3 Chapters from India is highly appreciated. In this book there are 4 more chapters: an introductory chapter concerning Smart City from Conasense Perspective and the establishing of the Conasense Branches in China and India, one chapter giving interesting visions and opinions on the intriguing question "What makes Cities Bloom and Prosper?" with key answer "Connected & Cooperating People", one chapter on recent results and prospects coming from future research showing the importance of "Human in the Loop" in smart vehicular systems, and finally a Chapter on "Business Model Ecosystem".

Seeing the outcome of this book, we come to the conclusions:

1. All efforts from the authors are highly appreciated;
2. The content of the book is important for local and national governments and for international governing bodies, for smart industries and for social organizations representing the interest of citizens and
3. The book is a success for Conasense and we hope that many people will read the chapters of this book.

So, we end this foreword on a simple note: we wish the reader to have much pleasure in reading and gaining new insights and ideas on Smart City from this book.

Leo P. Ligthart
Chairman Conasense

Ramejee Prasad
Founder member of CONASENSE

# List of Figures

# List of Tables

# List of Abbreviations

| | |
|---|---|
| A&OE | Administrative and Maintenance Expenditure |
| AI | Artificial Intelligence |
| AIDS | Acquired Immune Deficiency Syndrome |
| APP | Application |
| AVS | Autonomous Vehicle System |
| BCC Body | Coupled Communication |
| BI | Business Intelligence |
| BIS | Bureau of Indian Standards |
| BIS | Business Innovation & Skill |
| BRTS | Bus Rapid Transit system |
| btwieners | betweeners , weak links |
| CCTV | China Central Television |
| CCTV | Closed Circuit Tele vision |
| CEO | Chief Executive Officer |
| CETC | China Electronics Technology Group Corporation |
| CIDCO | City & Industrial Development Corporation of Maharashtra |
| $CO_2$ | Carbon Dioxide |
| CORS | Continuously Operating Reference System |
| CPSs | Cyber Physical Systems |
| CSS | Centrally Sponsored Scheme |
| CT | Computed Tomography |
| CTIF | Center for Teleinfrastruktur |
| CTSS | economic cooperation pact of the government of Turkey |
| CVDS | Cardiovascular Diseases |
| DMIC | Delhi–Mumbai Industrial Corridor |
| ECG | ElectroCardioGram |
| EHR | Electronic health records |
| email | electronic mail |
| EPC | Engineering, Procurement and Construction |
| ESB | Enterprise Service Bus |

| ETA | Estimated Time of Arrival |
|---|---|
| EU | European Union |
| Gaia | the name of godess Mother Earth |
| GDP | Gross Domestic Product |
| GIFT City | Gandhinagar International Finance & Technology City |
| GIS | Geographic Information System |
| GPRS | General Packet Radio Services |
| GPS | Global Positioning System |
| Ha | Hectare |
| HiL | Human-in-the-Loop |
| HMM | Hidden Markov Model |
| HTML | HyperText Mark-up Language |
| IaaS | Infrastructure-as-a-Service |
| ICF | Intelligent Community Forum |
| ICT | Information and Communication Technology |
| IOE | Internet of Everything |
| IoT | Internet of Things |
| IP | Internet Protocol |
| IPTV | Internet Protocol Television |
| IRL | In Real Life |
| IT | Information Technology |
| ITS | Intelligent Transportation System |
| ITU | International Telecommunication Union |
| ITU-TFG-SSC | International Tele Communication Union Focus Group Smart Sustainable Cities |
| JNNURM | Jawaharlal Nehru National Urban Renewal Mission |
| JUSCO | Jamshedpur Urban Services Corporation |
| LED | Light Emitting Diode |
| LOFAR | Low Frequency Array Radio telescope project |
| LTE | Long Term Evolution |
| M2M | Machine to machine |
| MB | Mega Byte |
| MIP | Mobile Instant Pages |
| MoUD | Ministry of Urban Development |
| MRI | Magnetic resonance imaging |
| MW | Mega Watt |
| NCT | National Capital Territory |

| | |
|---|---|
| NGO | Non-Government Organization |
| NIIF | National Investment & Infrastructure Fund |
| NIST | National Institute of Standards and Technology |
| NOC | No Objection Certificate |
| O2O | Online To Offline |
| OBOR | "One Belt One Road" economic pact, investment and construction project of the government of the Republic of China |
| OD | Origin Destination |
| OTT | Over-the-top |
| P2P | Peer-to-Peer |
| PaaS | Platform-as-a-Service |
| PB | Peta Byte |
| PCA | Principal Component Analysis |
| POS | Point of Sale |
| PPP | Public Private Partnership |
| PPT | People, Processes, and Technology |
| RFID | Radio Frequency Identification |
| SAAS | Software-as-a-Service |
| SASAC | State Owned Assets Supervision and Administration |
| SCADA | Supervisory Control and Data Acquisition |
| SCC | Smart City Council |
| SCP | Smart City Plan |
| SDM | Supervised Descent Method |
| SPV | Special Purpose Vehicle |
| SSC | Smart Sustainable City |
| SVM | Support Vector Machine |
| TB | Tera Byte |
| TOCC | Traffic Operation Control Center |
| TOD | Transit Oriented Development |
| UAV | Unmanned Aerial Vehicle |
| U-City | Ubiquitous City |
| ULB | Urban Local Body |
| UN | United Nations |
| UPS | United Parcel Service |
| UT | Union Territory |
| VCPS | Vehicular Cyber-Physical System |
| WBAN | Wireless Body Area Network |

| WHO | World Health Organization |
| Wi-Fi | Wireless Fidelity |
| WiMAX | Worldwide Interoperability for Microwave Access |
| WWW | World Wide Web |
| XML | eXtensible Markup Language |

# 1

# Smart City from Conasense Perspective

Leo P. Ligthart* and Ramjee Prasad**

*Founding Chairman Conasense
**Department of Business Development and Technology,
Aarhus University, Herning, Denmark

## Abstract

One of the biggest challenges in the world is to overcome major problems related to fast growing cities. Just using the expression "Smart City" is insufficient. In the last 10 years there have been many initiatives executed on what regional, national and international governing bodies called Smart City Construction. From experiences we learned that nowadays many renewed actions with large investments on Smart City Programs and Projects are needed. As a matter of fact, the term "Smart City" is often misinterpreted and its meaning is "enlarged" to include things that have not much to do with the term "smart", which is about including intelligence in systems and processes. To include intelligence, one needs to "sense", communicate (often remotely), process the sensed data to gain awareness and take decisions according to it. All these aspects are included in the Conasense concept.

This chapter is divided into three sections. Section 1.1 is the introduction on Smart City partly general but partly dedicated on Smart City from a Conasense Perspective. Examples as problems related to a Green Smart City with stringent requirements on air quality, and green transport without traffic congestions do not have solutions yet and, therefore ask for new and fundamental approaches. In this section special attention is given to recently started plans and projects on Smart City in China and India, two countries who face daily the benefits but also the immense problems connected with fast city developments. Section 1.2 of this chapter concerns the establishing of two new Conasense branches in China and India in 2016 and gives an overview of their activities. Both branches have given full support on the needed

*Breakthroughs in Smart City Implementation,* 1–18.

"Brain storm" for finding new routes which should lead to real solutions of existing City problems. The third subject gives a summary of all other chapters of this book.

**Keywords:** Smart City, Conasense Branches in China and India.

## 1.1 Introduction

More than 10 years ago we read much about smart cities. The fact that this subject still gets attention all over the world justifies the statement: Perhaps cities are not smart at present and another statement: Perhaps the existing problems of cities are not fully solved in 2025. In this respect it is very correct that the Brain tank "Conasense" sheds light upon problems and solutions connected to fast developing cities all over the world. There are many countries which have Smart City programs. In some cases the programs have been set up for relatively small cities with less than 500.000 inhabitants, other programs for medium-sized cities between half million and 2 to 5 million inhabitants and again other programs concern big cities between 5 and 20 million inhabitants. For example, The Netherlands started several Smart City programs in the beginning of this century; initiatives as the Delft Smart City program where Delft has only 100.000 inhabitants (really a small city) and the Amsterdam Smart City program where Amsterdam as the biggest city in The Netherland has less than 1 million inhabitants. Most European countries have several medium-sized cities and a limited number of big cities. Large number of big cities can be found in Asia.

In this respect a most interesting report on Smart Cities entitled "The development of smart cities in China" from the University Utrecht, The Netherlands appeared in 2015 [1].

The report describes case studies of Smart Cities in China. Visions and conclusions on these case studies are:

- Developing smart city at present goes primarily along two dimensions: government and market and not so much involvement of organizations a. representing "City Health and Well-being" such as hospitals, environmental institutes, etc and b. city inhabitants who notice marginal influence while experiencing not only the positive effects of cities such as housing problems, security, air quality, etc
- Telecommunication services providers benefit significantly from smart city initiatives. They expand their business areas and add to their network additional sensor and so-called "open platform" layers using cloud

computing technology and "Internet of Things" (IoT). The open platform allows third parties to supplement in technology and facilities and to develop Smart City applications.

- Constructing future smart cities will become more market-oriented and the government will focus more on standardization, law making, planning and comprehensive arrangements for all city plans and all citizens.

The report attracted our attention and in the Conasense Workshop 2016 several on-going Smart City initiatives in Europe, America and Asia have been overviewed. The outcome was clear: Conasense should be a platform for bringing forward our medium-term and long-term vision on solving Smart City problems which cannot be solved in short-term up to 2020. Based on above considerations the decisions in 2016 were:

1. Conasense book 2017 should be on Smart City and
2. The book is important for decision makers in

    a. governments for settling the long-term arrangements
    b. universities for solving fundamental problems which necessitate long-term research
    c. industry for future networks (more fail-safe technology, systems, and facilities and more wideband capacity per user)
    d. Institutes and citizen organizations for setting up the long-term requirements for Smart City where the Citizen has highest priority and where New demands on Quality of Life and Happiness of Citizen may lead to a different way on how all activities in cities should be organized in long-term.

This book aims to give an insight in on-going Smart City programs and projects in China and India, two countries where the number of big cities is larger than in other parts of the world. From all on-going research and actual programs and projects, together with the enormous series of articles in literature and the substantial list of actions in countries with smart cities programs, it becomes clear that there is ample proof for the statement "smart medium-sized and big cities do not really exist" and most medium-sized and big cities face enormous problems. Consequence of this statement is that urban developments did not automatically contribute to smartness of cities. We know that urbanization went fast in many parts of all continents and to the author's opinion it went too fast. Due to economic growth in and around cities it is understandable that in particular in developing countries many people from countryside moved to the cities.

This migration was facilitated by decision makers in politics, industry and social organizations. In that time urbanization was judged as positive for development of countries, regions and cities. As negative side of urbanization we can find many reports describing the impact that villages and small cities become smaller with fewer prospects for their future developments.

Lesson learned: starting in the last decennium of the 20<sup>th</sup> century signs appear that there are also negative aspects in too fast large city developments such as: not sufficient vacancies for work and big housing problems for all people who moved into the cities, lack of supportive infrastructure in and around the cities, and more. Cities were confronted with unbalanced developments between rich and poor, in employment for all new citizen, too expensive housing and too much transport by roads and rails. In the nineteen nineties it was invented that all city developments were not so smart. It explains why since that time smart city programs are being promoted. It is fascinating to note that in the last decade more and more actions are initiated to solve the unwanted problems in the cities.

Question is: "what makes a city a smart city?"

In many national and international initiatives, programs and projects we see that smart city of the future has a most advanced and not-yet existing infrastructure with smart buildings, smart roads and trains, smart cars and more using latest information and communications technologies (ICT) for support of new needed communications, navigation and sensing services. Of course in smart cities not only just mentioned infrastructure should become smart, but also major public and non-public developments are required to realize smart houses, smart industry and smart governmental and semi-governmental organizations. All initiatives on smart cities become less futuristic thanks to the enormous breakthroughs in ICT with "Internet of Everything" (IoE) and "Information Processing" (InfPro) and "Robotics".

Time is ready now that these breakthroughs allow for more interesting and challenging research and for developing demonstrators at a scale not shown before. But is this all?

The answer should be no. Via a questionnaire distributed among citizens in many European countries, it was asked what they consider as most important in life. The ranking of the answers representing the opinions of individuals is clear: First, "good private relations in family and nearby relations" and second "health and well-being". In other words Quality of Life (QoL) is seen as most important.

For the authors the outcome gives a clear sign: Smart cities should first of all directed to the "happiness in life of people" and "Smart city necessitates facilities" to improve QoL and to optimize the private and working environments in particular. Smart cities contribute to this happiness if not only attention is paid on smart roads and cars, but also major actions are undertaken to QoL privately, clean and healthy air in and outside the houses and smart working environment with flexibility of work in time and place.

This introductory chapter is written from the conviction that Conasense comes forward with visions how to contribute to the change going from "big city" to "smart city". This chapter is divided into three sections. After this Introduction, Section 1.2 describes the recently established Conasense branches in respectively India and China. Their smart city initiatives in both countries with highest population and many big cities under development deserve ample attention. Section 1.2 makes clear: cities become smarter by smart integration of actual Conasense topics related to Quality of Life and to novelties in ICT systems, networks and applications. Chapters 2 to 6 and Chapter 9 describe the present and future of Smart city developments in China and India.

Chapters 7, 8 and 10 of the book show that the "Brain Tank" Conasense is multi-disciplinary and that ample attention is paid on visions and ideas related to non-technological aspects which should be considered in Smart City programs.

In Chapter 7 we are informed about the "Human in the Loop" as part of Vehicular Cyber-Physical System (VCPS). It is appreciated that the authors paid so much attention to the uncertainty in VCPS coming from human behavior and human emotional states.

In Chapter 8 Professor J van Till makes us aware that at present most city tasks are strongly affected hundreds of millions of daily parallel messages and interactions.

The author stipulates that executing city tasks is not ICT technology only and gives us the lesson: "What individuals and groups of people do together with those power tools that matter". Chapter 10 motivates why the use of Business Model Ecosystems (BMES) is important in relation to future Smart Cities developments and that for better understanding of Smart Cities the BMES approach is valuable.

In Section 1.3, the authors give their conclusions of this chapter.

## 1.2 Establishing Conasense Branches in India and in China

Characteristic for Conasense is "brainstorming" on new technology and systems needed to satisfy not-yet existing services requirements. Visions and related roadmaps make clear that the requirements necessitate extension of available and/or recently developed ICT, particularly the extension of

a. Technology:

- Communications technology based on 5G and allowing for wide band and ultra-wide band communications in near ranges
- Internet of Things evaluating into Internet of Everything

b. Integration of existing and new circuits and devices leading into systems specifically developed for novel Conasense services

### 1.2.1 Conasense Branch India (CBI)

Learning from past experiences when developing new services, Conasense wants to take standardization seriously into account. The Global ICT Standardization Forum for India (hereinafter "GISFI") is an Indian standardization body active in the area of Information and Communication Technologies (ICT) and related application areas, such as energy, telemedicine, wireless robotics, and Biotechnology. India plays an increasingly important role in the processes of globalization in the field of the communications technology industry and research and development. GISFI is an effort to create a new coherence and strengthen the role of India in the world standardization process by mapping the achievements in ICT in India to the global standardization trends.

Further, GISFI is focused on

1. Strengthening the ties among leading and emerging scholars and institutions in India and the world
2. Developing and cultivating a research and development agenda for the field

Activities in GISFI fit well in the visionary plans of Conasense. GISFI has shown interest in Conasense since 2012. It was, therefore, a logic step to establish a Memorandom of Understanding between Conasense and GISFI. The intention of both parties is to co-operate for their mutual benefit in a broad range of multi-disciplinary areas focusing on future applications with strategic impact for Society. The collaboration facilitates the progress of the ICT industry allowing for breakthroughs in technological and application-oriented system developments needed to solve societal problems.

Signing MoU between Conasense and GISFI, May 2016.
Left: Prof R. Prasad, Founder Chair GISFI; Right: Prof. L. Ligthart, Chairman Conasense.

Conasense members witness the signing of the MoU between Conasense and GISFI.

In India there are several research-oriented universities and institutes as well as innovative ICT industries interested in the Conasense program.

Conasense is a global "think-tank" and already in the initial phase of Conasense it has been decided that in such a think-tank not more than 20 to 25 members coming from different countries should participate in Conasense Workshops. By this decision the meetings can be most effective with visionary presentations and in-depth discussions on future technologies and services.

Consequence has been that no more than 2 to 3 members from India can participate in the Workshop. In India the number of interested persons and organizations is substantially higher. A solution has been found by setting up a specific Conasense Branch in India, abbreviated by CBI and chaired by GISFI. Because Professor R. Prasad is Founder Chairman of GISFI and member of Conasense, he is elected to be the chairman of CBI.

Subjects to be handled by CBI in India should be coordinated with subjects in the global Conasense organization.

The MoU between Conasense and GISFI has opened the way to further establishing CBI. Official start of CBI has been January 1, 2017.

With broad support from the Conasense Board and from all other Conasense partners, the first subject in CBI concerns Smart Cities and CBI has been responsible for 3 related Chapters in this book. 5G is a central topic of research in India and, with its 1 Tbps data speed assurance, 5G is about to emerge as a promising way to strengthen the adherence among the three worlds of communication, navigation and sensing. Therefore, the 26th GISFI (Global ICT Standardization Forum for India) Standardization Series Meeting (GSSM), hold in Vishwaniketan, Navi-Mumbai, India, on September 26–27 2016, was dedicated to provide a platform in realizing CONASENSE in the 5G arena, with special focus on smart cities development [2–3]. Chapters 5 and 6 have been devoted to this issue by Indian authors.

Exchange of awards.

L2R: Vishwaniketan's President/GISFI Chairman, IETE Former President and IETE President.

Vishwaniketan Trusty Board Members.

Vishwaniketan's Director and Former IETE President.

## 1.2.2 Conasense Branch China (CBC)

In this century the country with fast and amazing developments is China in relative and absolute scales. China's Society experiences big changes in a multitude of social, economic and technological areas. For Conasense it is most interesting to have besides the Conasense Branch India also the Conasense Branch China. Despite China is extraordinarily active as developing country, at the moment China is already the second economic power in the world with ambitions to become the largest. China has a strong scientific community which wants to become a major player at world-level in all areas. As Adjunct-Professor at Beijing Institute of Technology (BIT), the highly-ranked university in China with top academic research on ICT and Remote Sensing, the Conasense Chairman had the opportunity to become acquainted with various ambitious and Conasense-related projects with strong support from China's government. His experience in ICT and Remote Sensing allows for transferring his system knowledge and participating in projects on Space-Based, Air-Borne and Ground-Based Atmospheric Remote Sensing and Earth Observation. In nearly all cases the BIT projects are part of collaboration programs with The Chinese Academy of Science, major national institutes as CETC and CASIC, ICT and Remote Sensing industries and several other universities. Already in 2013 BIT has strong interest to be linked with the "Brain Tank" Conasense because an organization as Conasense does not exist in China. BIT wants a key role in China on Conasense: "Communications, Navigation, Sensing and Services" and coordinator and stimulator of Conasense-related collaborative research in China. In the years 2014 to 2016 the BIT management and the management of The School of Information and Electronics in particular have used their network and contacts and investigated which ministries, institutes, and universities are interested in joining Conasense. The positive responses from many organizations made it worthwhile to set up in China the Conasense Branch China (CBC). In the same period there were many presentations on Conasense in China for The Ministry of Industry and Information Technology, two largest Chinese institutes CETC and CASIC, the Meteorology Institute focusing on space-based atmospheric science and technology, and several universities.

Meeting at the Ministry of Industry and Information Technology, April 2016.
Right: Prof. W. Hu.

Coordinator of CBC in China is BIT, Chairman the Dean of the School of Information and Electronics Professor Jianping An and Secretary Professor Weidong Hu. In the Conasense global Workshop CBC will be represented by two participants.

The Memorandum of Understanding (MoU) between Conasense and BIT was signed on November 10, 2016 by Professor An and the Chairman Conasense, witnessed by the BIT School of Information and Electronics management and supported by several Chinese organizations which become partner in CBC.

Signing the MoU between Conasense and BIT, November 2016.
Left: Prof. J. An, Dean School of Information and Electronics;
Right: Prof. L. Ligthart, Chairman Conasense

Vice-Dean Research and Vice-Dean Education of School of Information and Electronics and Prof. W. Hu witness the signing of the MoU between Conasense and BIT

In accordance with the prime characteristic of Conasense also CBC wants to be the organization in China where visions on future programs are developed with large impact on Society in 2025 and beyond. After ample deliberating upon the first area in which CBC will merge all needed expertise for a nationwide breakthrough with positive impact on Society in coming years but lasting for much longer time and certainly beyond 2025 a choice was made: Smart City. Main reason for this choice has been that China has most cities with highest population in the world and due to the growth in the number of inhabitants per city Smart City plans and programs are part of the inevitable trend of urban development in China. It is noted that Professor Hu has spent much time in the introductory phase of this first focus of CBC, entitled: "Concepts for Supporting Technologies and Applications of Smart City". In CBC the following considerations for breakthrough have been made: Services in Smart Cities concentrate on Smart Education, Smart Community, Smart Medical, Smart Grid, Smart Transportation and Smart Government Services. Priority in CBC for Applications is on ICT for Smart Transportation and Smart Infrastructure. New applications request for novelties in advanced technologies as needed to satisfy most demanding requirements for cloud computing, mass data communications, big data handling, on-line navigation, Internet of Things, Virtual Reality, 3D imaging, miniaturized sensors and radar, RFID's which should be realized in the period up to 2025. CBC aims to enter an era for China by design of new Smart Cities using latest technologies with new insights on complex systems control and allowing for functioning 24 hours all days. Integration and implementation of these visions and technologies may facilitate new services, efficient management support, development on improving city environment including air quality, information sharing using secured and standardized networks and many more.

In 2015 and 2016 the chairman of Conasense launched the idea to include in the Chinese program on Smart cities a focused program on improving air quality. The importance of good air quality in big cities is urgent worldwide and in a city as Beijing air quality has big effect on health and economy as we learned from own experience. In Conasense ideas are discussed how to develop air cleaner which can in future vacuum up the polluted air, filter the air and bring the cleaned air back in the atmosphere. This visionary and exciting plan may be selected in a future Conasense program so that it will play a significant role on two key issues important in the Conasense mission: contributing to

1. Future "health and well-being in cities" at world level scale
2. "Happiness in life" of people.

In this book three chapters describe the present and near future of Smart City in China. Two chapters are from BIT (authors W. Hu and J. An); the third chapter from CETC entitled "Smart Cities in China" contains background of on-going activities started in last years. It starts by motivating why Smart City development is urgently needed.

Main reasons are:

1. Serious urban diseases exist such as population expansion, resource shortage, serious environmental pollution, traffic congestion and increasing public safety hazards
2. Smart city ideas based on the new generation of information technology bring a new urban life expectation for government and people.

Concepts and characteristics of smart city are given, followed by "why and how" the constructing of smart cities in China is a major strategy in economic and developmental transformation, and is an important step towards an innovation-oriented country. The remaining part of the chapter gives details on sub-systems and key technologies of the integrated smart city system.

The second chapter "Intelligent Services Supporting On-Going and Future Urban Developments in China" starts from the decision of Chinese government in 2013 to support 193 pilot projects on smart cities in China in 2013 for generating new possibilities of industrialization, informatization and urbanization. Attention is paid to the roles and relationships of various actors (government, market, and society) in smart city development. It expounds features of Smart Cities and demonstrates the paths for technology and application developments characteristic for cities all over China. This chapter ends with experiences and visions on future developments.

## 1.3 Conclusions

Smart City initiatives in Europe, America and Asia made clear that Conasense should present her medium-term and long-term vision on solving Smart City problems which will not be solved in short-term up to 2020. We decided that the Conasense book 2017 should be on Smart City with visions and opinions important for decision makers in governments, universities, industry, institutes and citizen organizations.

Starting from the idea that at present "smart medium-sized and big cities do not really exist" and "urban developments did not automatically contribute to smartness of cities".

We explain in this chapter some negative aspects in too fast large city developments and we note that actions are initiated to solve unwanted problems in new cities.

We see initiatives for interesting and challenging research and for developing demonstrators at a scale not shown before. But in Conasense perspective this is not sufficient. Equally important is "happiness in life of people living in cities" and "Smart city necessitates facilities to improve Quality of Life". In summary, we promote not only attention for smart roads and cars, but also for clean and healthy air in and outside the houses and smart working environment with flexibility of work in time and place. From this perspective, technology is extremely important for fulfilling the requirements of smart cities, but it is not enough and a more cross-disciplinary effort, as promoted by Conasense, is needed.

The 2 newly established Conasense branches in China and India have played an important role why the topic of Smart City is put so high on the Conasense agenda. In Section 1.2 of this chapter the full background of both branches has been presented.

This introductory chapter is written from the conviction that Conasense comes forward with visions how to contribute to the change going from "big city" to "smart city".

## References

[1] Yongling Li, Yanliu Lin and Stan Geertman, *"The development of smart cities in China"*, University Utrecht, 2015.
[2] http://www.gisfi.org
[3] http://vishwaniketan.edu.in

## Biographies

**Prof. Dr. ir. Leo P. Ligthart** was born in Rotterdam, the Netherlands, on September 15, 1946. He received an Engineer's degree (cum laude) and a Doctor of Technology degree from Delft University of Technology. He is Fellow of IET and IEEE and Academician of the Russian Academy of Transport.

He received Honorary Doctorates at MSTUCA in Moscow, Tomsk State University, and MTA Romania. Since 1988, he held a chair on Microwave transmission, remote sensing, radar and positioning and navigation at Delft University of Technology. He supervised over 50 Ph.Ds.

He founded the International Research Centre for Telecommunications and Radar (IRCTR) at Delft University. He is a founding member of the EuMA, chaired the first EuMW in 1998, and initiated the EuRAD conference in 2004.

Currently, he is emeritus professor of Delft University, guest professor at Universities in Indonesia and China, Chairman of CONASENSE, Member BoG of IEEE-AESS.

His areas of specialization include antennas and propagation, radar and remote sensing, satellite, mobile, and radio communications. He gives various courses on radar, remote sensing, and antennas. He has published over 600 papers, various book chapters, 4 Patents and 6 books.

**Dr. Ramjee Prasad** is a Professor of Future Technologies for Business Ecosystem Innovation (FT4BI) in the Department of Business Development and Technology, Aarhus University, Denmark. He is the Founder President of the CTIF Global Capsule (CGC). He is also the Founder Chairman of the Global ICT Standardisation Forum for India, established in 2009. GISFI has the purpose of increasing of the collaboration between European, Indian, Japanese, North-American and other worldwide standardization activities in the area of Information and Communication Technology (ICT) and related application areas.

He has been honored by the University of Rome "Tor Vergata", Italy as a Distinguished Professor of the Department of Clinical Sciences and Translational Medicine on March 15, 2016. He is Honorary Professor of University of Cape Town, South Africa, and University of KwaZulu-Natal, South Africa.

He has received Ridderkorset af Dannebrogordenen (Knight of the Dannebrog) in 2010 from the Danish Queen for the internationalization of top-class telecommunication research and education.

He has received several international awards such as: IEEE Communications Society Wireless Communications Technical Committee Recognition Award in 2003 for making contribution in the field of "Personal, Wireless and Mobile Systems and Networks", Telenor's Research Award in 2005 for impressive merits, both academic and organizational within the field of wireless and personal communication, 2014 IEEE AESS Outstanding Organizational Leadership Award for: "Organizational Leadership in developing and globalizing the CTIF (Center for TeleInFrastruktur) Research Network", and so on.

He has been Project Coordinator of several EC projects namely, MAGNET, MAGNET Beyond, eWALL and so on.

He has published more than 30 books, 1000 plus journal and conference publications, more than 15 patents, over 100 Ph.D. Graduates and larger number of Masters (over 250). Several of his students are today worldwide telecommunication leaders themselves.

# 2

# Construction of Novel Smart City in China

Menglan Jiang and Guizhong Xu

Institute of Science and Academy of China Electronics Technology,
Group Corporation, Beijing, China

## Abstract

Concepts of digital city, intelligent city, and smart city play an important role in the integration of information, industrialization, and urbanization. However, most of the larger cities in the world are facing serious problems, such as climate, environment, food safety, energy, transportation, and public safety. The economic crisis has led to a decline in urban quality with the consequence that urban management and industrial development gradually lost its way. In this context, the next generation of Internet technology acts as an extra driving force which has been brought into action in order to solve existing problems and contributed to providing a possible vision and program for the future. Smart city will not only change the way of life of residents, but also change the mode of urban development with an accent on sustainability and intelligence for the city in the future.

**Keywords:** Smart City, Industrial Upgrading, National Strategy.

## 2.1 Main Situation of Smart City

China began to accept the concept of smart city in 2010 and promoted this concept for cities varying from large cities to medium-sized cities. According to statistics, until June 2016, over 95% of all cities in China are provincial cities while more than 76% belong to small cities. More than 500 cities have clearly put forward the concept of smart city and built the construction of smart cities; in this sense China has become the test bed of world's smart city constructions. However, there is no thorough study of the concept of smart

*Breakthroughs in Smart City Implementation,* 19–36.

city and also the social benefits are not obvious, especially in the field of e-government and public services [1].

With the fast development of urbanization, the problems of the city become more and more prominent; the construction of smart city has become the forefront trend in the world. China has entered the phase to bring the smart city construction into practice; the program is called "the new national planning (2014–2020)" in which the idea of "promoting the construction of smart cities" is explicitly proposed. Each Chinese ministry has released a number of policies for supporting the idea. With the approval of the state council in 2014, the national development and reform commission, the ministry of industry information and the ministry of science and technology jointly formulated "the guidance of promoting healthy development of smart cities". Smart city construction has been formally incorporated into the national industry development strategy.

In August 2015, the "Action Program of Promoting the Development of Big Data" has been promulgated by the state council. It clearly put forward the main objectives directed to building

1. new model for social management allowing for more precise management and collaborative;
2. new economic operation mechanism allowing for more smooth, safe, and efficient operations;
3. new people-oriented service system; and
4. new set-up for public entrepreneurship and people innovation.

In December 2015, the nation-wide "city work conference" has been held. "City Work" aims to increase the level of "city work" over the country with the intension to organize a major change in city work. In March 2016, the idea of "building a batch of novel smart cities" has been proposed. In April 2016, President Xi Jinping stressed the importance of

1. Promoting the modernization of national management system and management abilities via informatization;
2. Hierarchical promoting the construction of smart city;
3. Lowering information barriers and building a national information resource-sharing system;
4. Using information methods to become more aware of the social situation;
5. Clearing communication channels; and
6. Auxiliary decision making.

In the national 13th five-year information planning, the building of a unified open "Big-data" system and carrying out the construction actions for new smart cities have been proposed. In general, the field of novel smart city will meet a set of challenges under the support of national strategies during the 13th five-year planning. Terms of smart government, smart city construction has undergone three stages of development from government office automation and online government information to e-government. Goals of smart government include integrating resources, serving the people, synchronizing approvals, creating a transparent government, and achieving smart decision-making.

## 2.2  From Smart City to Novel Smart City

China has already received significant achievements in the construction of smart city which has also been affirmed by the rest of the world. However, for the development of urbanization, it will bring more challenges and pressure to the urban management, construction and development. It is understood that until the year 2030, the population due to urbanization will increase from 760 million to 900 million. How to deal with the great challenge of future urban developments has become the common problem for the whole world. In this background, the idea of novel smart city has been proposed based on theory and practice of previous smart city results [2].

As a new stage of smart city development, the novel smart city is the result of a strong fusion between the modern information technologies and steps in urbanization. "New" in novel smart city concerns three different aspects which are:

- Breaking the information "funnel" to realize information interconnections.
- Realizing cross-industry big data integration and sharing.
- Building a city information security system to ensure the safety of the city [3].

The traditional smart city mostly focus on technical levels, that is, infrastructure network, sensing devices, cloud computing, universal platform, and basic information resources, etc. The novel smart city implements more modern management, smarter operations, safer development, and happier life of people by fusing of

- information planning;
- leading and pushing a new generation of information technology; and
- city modernization and iterative evolution.

In general, the main contribution of smart city is improving the abilities of government and the social management system to provide better services to the people [4].

From smart city to novel smart city is in fact the evolution and iterative process from a level 1.0–2.0. The traditional smart city (or city 1.0) can be considered as the early stage of the construction of smart city. At this stage, information and technology have been emphasized. City information construction can be implemented via different types of information technology, urban management, public services, and industry development; examples are electronic government actions and the information system construction. However with the improvement of different types of the information infrastructure, the concept of smart city is becoming more and more mature. Big data, cloud computing, Internet-of-Things, mobile Internet and artificial intelligence are rapidly developing. Only focusing on the information construction is not enough for satisfying the future long-term and sustainable requirements, while the traditional smart city construction generates problems such as the so-called "information chimneys" and "data islands". Therefore, the evolution from traditional smart city to the novel smart city has been considered as the inevitable trend.

Compared with the traditional smart city, the novel smart city should still be based on all types of information infrastructure and focus on information sharing, big data mining and utilization, and city safety. Key points of novel smart city construction are:

- Breaking the information and data islands of traditional smart city;
- Implementing data collection, sharing and utilization; and
- Establishing a unified big data operation platform.

These points can play a significant role in initiatives on "being a nice government," for "better life of the people" and for "improving industry conditions." Meanwhile, with the improvement of city information and intelligence, the problem of information security has received more attention in novel smart city construction aiming to ensure the safety of all kinds of city information and security of big data. Finally, the city development aims ultimately to serve people and to promote a better life for people. Therefore, the novel smart city makes "information security" as a starting point, and will be people-oriented as a most significant feature for the novel smart city.

From the time the idea of the novel smart city has been proposed, Chinese government provides great support and active guiding from the policy level. In March 2016, the idea of "making infrastructure intelligent, better public service facilitation, and improving social management as key points for building a batch of novel smart cities and by using modern information technology and big data" has been strongly proposed in the 13th five-year plan. In April 2016, "The launch of 100 new smart cities is in the 13th five year plan" as proposed by the national development and reform commission. In November 2016, the annual development report of "The novel smart city development report 2015–2016" has been officially published. The report comprehensively introduced the developing idea, related work progress, and the latest research results of the novel smart city. Experiences from some outstanding cases have been discussed to promote novel smart city developments. On November 22nd, "The report about organizing novel smart city notice" has been published by the national development and reform commission, the central internet and information office, and the national standardization committee. The report states to begin with the evaluation work of the novel smart city according to the so-called "evaluation index" [5].

Intelligence is the inevitable trend for future city development; moreover, the size of the future smart city will increase as well. Meanwhile, with the development of all types of new ICT technologies, more and more ICT companies will join the workforce of the novel smart city. However, the novel smart city construction is a complex system project with many links, domains, and departments. These all decide that the construction of the novel smart city cannot be completed by a single government or enterprise, it can only be developed by both government and smart city companies to make our city smarter and to benefit everybody [6].

"Developing a batch of novel smart cities" has been introduced in the outline of the national 13th five-year plan; related departments would select 100 cities as "pilot" and begin their evaluation work for the novel smart city. In October 2016, Chinese chairman Xi Jinping pointed that "Building an integrated big data center per city allowing for cross-level, cross-regional, and cross-department business management and services" can further improve the development of the novel smart city [7].

## 2.3 The Developing Trend of Novel Smart City

### 2.3.1 Internet Enterprises to Participate in the Expansion of New Intelligent City Constructions

Applications and cloud computing have been considered as breakthrough for Internet companies and an open cooperation model is useful for promoting the novel smart city constructions. At the same time, China will promote companies to participate in the development of the novel smart city via buying services and government guidance. At present, characteristics of the novel smart city present cooperation rather than competition. The construction of novel smart city is a huge project which involves multiple levels in all participating organizations and needs cooperation among different manufacturers [8].

### 2.3.2 Government–Enterprise Collaboration for Gradually Replacing Government Investments

The novel smart city handles information on city developments at highest level of responsibility. It includes city development plans with all kinds of new elements and content. If the government focuses on both management and operations, it may easily lead to many problems such as insufficient urban finance, low sustainable development capability and low management efficiency. By interaction with investors having the power of private capital, with market mechanisms and business philosophy, the city management can expand the comprehensive management of urban resources and enhance the city management capacity and quality. It has been proven that government and enterprise cooperation rather than a purely government-based approach is more conducive to the implementation of novel smart city construction. Therefore, in the 13th five-year plan, this trend will be more explicitly amplified [9].

### 2.3.3 Big Data Mining Enhance the Novel Smart City Experience

In the smart city construction process, China has begun to use at a large-scale cameras, RFID, etc. in order to collect city traffic, security, flow, logistics and other aspects of information. City managers have begun to use this information to manage. However, due to technical constraints, the city data is not really efficient for use and the smart city did not achieve the level of "truly smart". The development of big data technology is to solve the city's "low IQ" problem. It should provide an effective tool for the construction of novel smart city and lay a solid foundation for "fully smart" technology and system in future. Some typical example applications of big data analysis in intelligent traffic systems can be found on Internet [10].

### 2.3.4 Information Security: Strategic Focus in Novel Smart City

At the first meeting of the Central Network Security and Information Leadership Group held in February 2014, General Secretary Xi Jinping specifically stressed that cyberspace should be secured. In the process of building a new intelligent city, infrastructure and information resources are important parts of its construction and will directly affect the effectiveness of the "new wisdom" of the city. Information security as an auxiliary support system is most important in the new generation of intelligent city construction. How to build an information security comprehensive monitoring platform and how to strengthen the information security risk assessment system, will become a strategic focus in the new smart city. As part of the 13th five-year plan, the government will focus on the classification of infrastructure and continue to strengthen the network infrastructure and information resources using security protection methods; enterprises should strengthen industrial cooperation, to form a joint force to promote the development of China's security information industry.

## 2.4 China Electronics Technology Group Corporation (CETC) Novel Smart City Construction Concept and Development Ideas (1)

### 2.4.1 Construction Concept (1)

As consequence of bringing novel smart city into practice, lots of companies have started their constructions, including CETC, Huawei, ZTE and many more. Being a national team and main force in the field of electronic information, CETC actively responds to the national development strategy, and has come up with a new strategy called "One Five Five Three", which mainly focuses on intelligence and security. The novel smart city is the priority among priorities in CETC intelligence activities. It is also the most incisively and vividly reflection of the five development ideas: Innovation, Harmony, Environmental, Open and Shared. The general interface of the smart city made by CETC is shown in Figure 2.1.

Under the guidance of our "innovation, harmony, environmental, open, and shared" development concept and the inspiration we got from the central city work conference, CETC makes best use of the in-house technology knowledge and top-level design experience, by presenting the "six ones" concept of the novel smart city:

**Figure 2.1**   General interface of the smart city.

- Build an open architecture. Novel smart city is a complex huge system, whose construction has to obey system rules with an open architecture allowing for building a system consisting of one part for integration of IoT basis facilities, a second part as a general function platform and a third part for smart city applications. It follows the principles of a system with a strengthened (robust), integrated and open architecture, to be used for a variety of smart cities. This approach makes the system transplantable, tailorable and interoperable.
- Build a general so-called "Basis Net". The network follows ideas on "space-based networking, ground net spanning generation and space–ground interconnection" by building a part of a space-ground integration city information service grating net. The net should support a large number of perception and computing (plug and play) devices with stabilized transmission. Purpose of the net is to promote usage intensification utilizing available resources of communication and computing, to realize city precise perception, to interconnect among information systems and to offer a ubiquitous people-benefit service.
- Build a general function platform. To enhance the compatibility of computing, storing, networking, exchanging objects and data resources by different manufactures, a general function platform should be built. In the platform the following functions are executed: dispatches administration and packaging of all kinds of information resources, boosts inner

city cross-district, cross-department, cross-layer application integration, supports the wisdom of city administration public service, and increases system service efficiency.

- Build a data system. The novel smart city needs to converge data from traffic, medical treatment, sanitation, industry, public security, agriculture and many other domains, which requires an open and shared data system. By reorganizing and fusing data, it forms a summation of data, and then increases the production and applying of data that effectively improves decision making and a more scientific and intelligent attitude of the city administration.
- Build a high-efficiency operation center. The operation process of the novel smart city is sensing and conducting video monitoring, utilizing an underground pipe network and other social information resources, coordinating every business department dispatching. Finally, it results into a more safe and social stable macro economy. It helps a better controlling and managing urban municipal facilities, public safety, eco-logical environment, macro-economic, and public opinion of the people's livelihood, share resources of gathering city and inter-departmental coordination linkage.
- Build a unified standardized system. In order to ensure that new orderly and healthy development of the novel smart city, government leading is needed and should be based on the city characteristics, classification of planning, construction content, and core elements. It establishes and improves the construction and evaluation of the standardized system on aspects as giving prominence to the overall standard, coordination and compatibility, and better support for top-level design and system construction.

## 2.4.2 Construction Concept (2)

### 2.4.2.1 The top-level design as core activity for building an open system architecture (Figure 2.2)

The speech on smart city construction by general secretary Xi, concentrates on the requirement of improving the capacity of the national governance system and management and the "innovation, coordination, green, open, sharing" development philosophy. Using operational system engineering methods, the key to strengthen the top-level design, is building an open architecture with an advanced novel smart city top-level design leading towards an architecture, processing, methods, standards and general resources capability allowing rapid promotion, and to be copied nationwide.

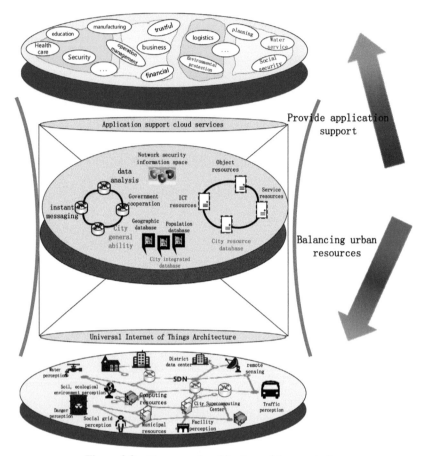

**Figure 2.2**   The general architecture of the smart city.

There are three different layers in the figure which are: smart application, general functional platform, and infrastructure. The main function of the general functional platform are providing support to the application layer and controlling available resources in the infrastructure layer.

### 2.4.2.2 Two-wheel drive technology and mechanism innovation

Using Internet, Internet-of-things, big data, cloud computing, and other advanced information technology can be fused into the New Smart City construction, improving the management ability of the city and the realization of better service for the people. Motto here is: give full play to the market mechanism, explore the PPP mode and service outsourcing, and promote the development of city management system reform, revitalizing the city data resources, and information resources.

### 2.4.2.3 Construction of open cities and support for economic development

Through the city operation management center and general function platform products new functionalities become possible to promote cross-industry interdisciplinary data fusion, to improve data value and build a data source platform, to attract domestic and foreign enterprise application development service, to support innovation entrepreneurship and to create an open city, digital economy and information economy and the "Internet+" industry developments.

### 2.4.2.4 Strengthen the network space safety management

By integration of cyberspace security solutions, the general architecture allows to build the network space safety overall capacity, to enhance the level of system security, data security and infrastructure security, and to resolve and relieve urban security problems in the process of informatization construction.

At the same time, three universities and 19 companies (domestic and abroad) jointly sponsor the activity through "the novel smart city" construction enterprise alliance. Meanwhile in July 2016, CETC, ISOFT stone, Taiji and other enterprises have set up a novel smart city ecosphere, published an ecosphere action plan, and are working together to build the novel smart city.

## 2.5 CETC Novel Smart City Construction Concept and Development Ideas (2)

With the support of Cyberspace Administration of China, National Development and Reform Commission, Ministry of Industry, and Information Technology and some other ministries, CETC signed the novel smart city strategy cooperation agreement with the city of Shenzhen, Fuzhou, and Jiaxing, respectively, during the Second World Internet Conference. CETC originated the novel smart city Enterprise alliance with more than 40 famous domestic and overseas companies. After the conference, CETC formed expert teams to conduct demanding research related to the three example cities, which focus on top level design, city operation management center construction, general function platform exploitation and novel smart city enterprise alliance. This work has got an outstanding effect.

### 2.5.1 The Construction of New Smart City Research Institute (1)

China Electronics Technology Group Corporation registered the "CETC Shenzhen Novel Smart City Research Institute co., LTD" and undertook the following actions on

- novel smart city development strategy,
- pooling CETC and global industry high-end knowledge,
- creating novel smart city strategic research, innovation, transformation, and operation of service highland,
- advancing new knowledge on city industry pattern formation and industry sector gathering, essential for the new smart Shenzhen, new smart Fuzhou, new smart Jiaxing, and communities; these actions are directed to the building of a novel smart city new standard and a new model, with exposure throughout the country, and
- building a domestic outstanding, world first-class novel smart city with an overall solution provider, operators and service providers.

China Electronics Technology Group Corporation positively cooperates with the Fuzhou and Jiaxing local governments, and with the enterprise alliance members, according to the thinking of "joint construction, joint operation", for planning, promoting and running the needed actions by the local authorities, and explores the new operational model and business model.

### 2.5.2 The Top Design Work of the New Smart City

For the implementation of the novel smart city construction "six ones" concept of CETC, it is necessary to make benchmarking demonstrations and to promote the pilot city novel smart city top level design work benchmarking. Since August 2015, CETC has started the research work in Shenzhen, Fuzhou, and Jiaxing with hundreds of departments of the three cities, operators, utilities organizations, and administrations in the region. Until now, they have completed the top level design plan of Shenzhen, Fuzhou and Jiaxing. On May 19, 2016 it passed the review by 6 academician of the Chinese Academy of Engineering, 12 experts in domestic city planning and construction, and in information technology. The result is considered to represent the current highest level of domestic smart city top level design, and has been complemented by the two municipal leaders' endorsement. Under the guidance of the top level design framework, the related departments have carried out the planning of Jiaxing for the three years rolling out program and for key projects, which has passed the Jiaxing city CPPCC review; the related so-called blueprint document describes the novel smart city construction of Jiaxing planning.

### 2.5.3 Construction of City Operation and Management Center

From the demands for a modernized and intelligent city management capability, CETC is building a strong and unified operation platform. It concerns the construction of an operation management center able to do complete fusion and presentation derived from government departments and from business data. It provides a time sensitive synthesized situation display and assists decision making application by five capabilities as city operation synthesized test, city operation synergy dispatch, city development plan as a whole, city data resource open service, and cyber space security administration. The city operation management center follows the important functions as major activity and separates the work into some stages. The logical architecture of the smart city based on the universal functional platform has been indicated in Figure 2.3.

## 2.6 The Construction of New Smart City Research Institute (2)

The novel smart city "General function platform" is a fundamental and framework platform for constructing and operating a novel smart city by CETC. Lying in the middle level of the city operation management central architecture, the platform plays a core role in connecting downstream integrating data and basis resource, and upstream supporting city application services. The platform synthesizes, improves and consummates system integration, data fusion, big data computation, cloud platform, city operation control, cyber security and many other key technology packs. At the moment (2017), we are undertaking the software exploitation and planning to finish the first vision product by the end of the year.

In connection with the starting of novel smart city top level design, CETC pushes forward the study of standards by three dimensions of technology, system mechanisms and operation evaluation. By negotiating with organizations CAC, NDRC and SAC, CETC became a member of the Smart City Standardization Working Group, and is involved in the work on evaluation of the index system and on drafting related standards.

To better converge intelligence worldwide and to push forward the construction of novel smart city, CETC has an alliance with more than 40 famous advanced technology companies, universities, and institutes organizing the

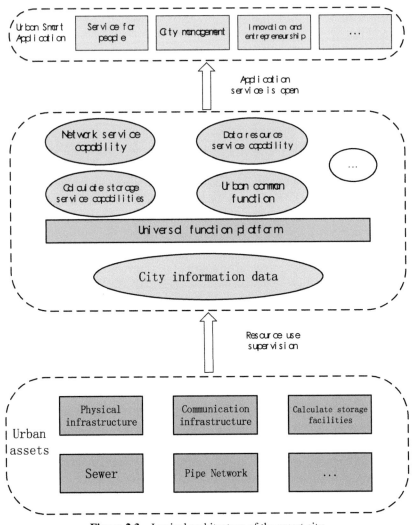

**Figure 2.3**    Logical architecture of the smart city.

"novel smart city construction enterprise alliance", which was announced at the Second World Internet Conference. Through this alliance the investigations are done in a cooperation mode and it can motivate the integration of the upstream and downstream industry chains, and build dispute synergy innovation platform to realize industry aggregation developments. The alliance secretariat has planned many meaningful activities, such as the visits among the alliance members. By "communication, cooperation, open source and joint sponsorship" the secretariat has founded a working mechanism, which is widely supported by the alliance members.

## 2.7 Conclusion

In China high priority is given to the construction of Novel Smart Cities starting from 2015. When Novel Smart Cities are operational it may have big impact on national and regional (and city) government, universities, institutes, industry and most important on happiness and quality of life for the citizens. Main reason why new concepts for smart cities are needed is the experience with earlier Smart City initiatives which was not only positive.

First phase of Novel Smart City design have been detailed in this chapter. CETC as a major research-oriented organization plays and will play an important role in the on-going and future actions.

## References

[1] http://3y.uu456.com/bp_1oatx0n6pz3gyk7183zq_1.html
[2] State Forestry Administration of the People's Republic of China. Available at: http://www.forestry.gov.cn/portal/xxb/s/2519/content-929417.html
[3] Fang, D. (2016). *Du Mingfang: Interpretation of China's New Wisdom City* Available at: http://mt.sohu.com/20161215/n475910296.shtml
[4] Chengdu Daily. (2016). *What is the new wisdom of the city?* Available at: http://www.sc.xinhuanet.com/cxjs/2016-03/14/c_1118323283.htm
[5] OFweek. (2016). Inventory: 2016 new smart city construction have those achievements? Available at: http://security.ofweek.com/2016-12/ART-510011-8420-30078313_2.html
[6] Zhong, D. (2016). What is the new "new" compared to the traditional wisdom of the city? Available at: http://www.forestry.gov.cn/portal/xxb/s/2519/content-929417.html
[7] (2016). 2016 new smart city construction with those achievements? Available at: http://mt.sohu.com/20161216/n476109919.shtml
[8] Zhongguancun Online. (2015). Wisdom city ten trend of Internet enterprises to participate in the construction. Available at: http://tech.hexun.com/2015-02-24/173501454.html

[9] Cnii. (2015). The development of smart city in 2015 will show five trends Available at: http://www.cnii.com.cn/thingsnet/2015-04/01/content_1554892.htm

[10] ICESbSHKBIM Institute. (2016). 2016 intelligent city construction to a new era. Available at: https://sanwen8.cn/p/107rfmX.html

## Biographies

**Menglan Jiang** was born in Heilongjiang, China. She received the B.S. degree in computer science from University of Electronic Science and Technology of China (UESTC), Chengdu, China, in 2007, the M.S. degree in information security from UESTC in 2010, and the Ph.D. degree in telecommunication from King's College London (KCL), London, UK, in 2016. Since 2017, she has been with the Institute of Science Academy of China Electronics Technology Group Corporation (CETC), Beijing, China, where she is currently the Technology & Standardization Specialist. Her current interests include the 5th Generation (5G) wireless network, resource management, machine learning, and big-data analysis.

**Guizhong Xu** (1983.09), received his master degree from Communication University of China, 2010. Now, he has been with the Institute of Science Academy of China Electronics Technology Group Corporation (CETC), Beijing, China. His main interests include new smart city construction, video intelligent processing, video retrieval (2D/3D), video encode and decode, video communication and Internet of things technology.

# 3

# Intelligent Services Supporting On-Going and Future Urban Developments in China

Weidong Hu and Jianping An

School of Information and Electronics, Beijing Institute of Technology

## Abstract

Since IBM brought the concept of "Smarter planet" in China in 2008, smart cities construction has become a new trend of urban development. By 2013, there have been 193 approved pilot projects of smart cities in China. Smart city is viewed as key strategy to promoting industrialization, informatization and urbanization. The rapid development of smart cities in China is largely attributed to the cooperation between IT companies and the government. This chapter will describe how smart cities have been developed in China. It will pay particularly attention to the roles and relationships of various actors (including the government, market and society) in the development of smart city. It is a remarkable issue that major cities in China are able to transform from information and digital technology city into a "Smart City"; also it is an inevitable requirement for major cities to adapt itself to more complicated management. This chapter expounds the features of Smart Cities based on the development in many cities and demonstrates the technological application development paths to be the smarter cities based on unity and diversity characteristic for cities all over China. This chapter will finally summarize Chinese experiences on the development of smart cities and give visions on future developments.

**Keywords:** Smart City, Intelligent Services, Information Technology.

*Breakthroughs in Smart City Implementation,* 37–86.

## 3.1 Introduction

Smart cities take full advantage of the benefits of information and technology. By monitoring, analysing, integrating and smart response, it is possible to optimize resource allocation and to enjoy the convenience and intelligence of resource allocation applications, so as to facilitate the residents who live in smart cities.

At present, if we examine the project details of smart city construction, we can see that an important aspect is strengthening urban communication and network infrastructure, i.e., to increase the coverage and bandwidth of communication network. Another important topic is providing intelligent application services in key areas, such as smart public services, smart social management, smart transportation, smart medical care, smart security systems, smart logistics, smart living, and so on. Among them, smart transportation, smart urban security systems, and smart medical care are key areas. The cities which meet the software and hardware foundation can begin to construct the smart cities. The research and practice of smart city construction in our country has just started, so there is no any basic mode for reference. In this chapter we describe the basic mode of constructing a smart city and provide a series of reference modes for common problems occurring in the upstart of smart cities. One mode is the basic target to compose an infrastructure layer, resource layer and application layer. The basic target mode is designed from research on programming the construction of a smart city and referring to related literature. Another is the implementation mode of smart city construction which includes the subjects of collaboration mode and the investment and financing mode. This implementation mode is based on research results for constructing the Chinese smart city. The third is the process management mode of the smart city construction which is based on public sector strategic management theory, organizational vision theory, harmony management theory and other ideas and methods. In this chapter we draw the structure maps of each mode.

## 3.2 Smart Government

In terms of smart government, smart city construction has undergone three stages of development from government office automation and online government information to e-government. Goals of smart government include integrating resources, serving the people, synchronizing approvals, creating a transparent government and achieving smart decision-making.

Smart government is generally divided into four parts. The first part is the public service data exchange platform which aims to gather basic government data scattered from various functional departments into the public government data centre and to create a public information database. Based on this, several thematic databases can be further established. The establishment of public databases can greatly facilitate urban residents, since they can search for relevant information such as personal status, registration status, medical accounts and social security accounts by simply checking their city cards. For government administrators, they can have more efficient and convenient access to various types of urban operations data such as demographic data, corporate data, and spatial data and so on.

Another prominent feature of smart government is the smart city grid, the construction of which can strengthen urban management. City grid is based on unified urban management and digital platforms that can divide urban area into unit grids according to certain standards. Through the monitoring and supervision of the grids, we can establish a management system that can separate supervision from treatment. This allows administrators to voluntarily detect and timely process, and thereby strengthens government's administration capabilities and processing speed, solving problems before they are complained by residents. Today's mobile government is one example of bringing convenience to people. It combines mobile communication technology with internet to break time and space constraints, enabling e-government to provide services without subject to time, place, forms or personnel.

The administrative service centre based on the public data centre and smart city grid is an indispensable part of smart government. All functions of smart government can be achieved on this platform.

Last but not least, the emergency command and operation system play an important role in dealing with major events, disasters and dangerous situations.

In short, smart government is essentially creating a digital, internet-based, and intelligent environment to perform various functions. These functions include government information disclosure, administrative licensing approval, social management, public service, performance monitoring, information sharing, and daily office routines. With the support of modern technologies such as computers, network communications, IoT and etc, communities and the public can enjoy smart services. Government can take advantage of many smart management tools.

### 3.2.1 Case Study: Beijing

In October 2010, Beijing launched Beijing Xiangyun Project Action Plan [1, 2], which aims to integrate important industrial factors of the cloud computing industry, including talents, venture capital, industrial bases, innovative enterprises and more. The action plan aims to promote the cloud computing industry to develop early and fast at large scale in order to seize the initiatives in the new round of global competition in information technology. The plan also hopes to drive the whole cloud computing industry chain to realize a 200-billion-yuan value. Since then, Beijing Xiangyun project has become the most compelling project among all the cloud computing pilot projects in China. Beijing Government has played a leading and exemplary role in the project. Of all kinds of government information systems, the government cloud computing platform is the system that demands the most of cloud computing.

Beijing government cloud platform has been further enhancing the capacity of information technology infrastructure with the help of cloud computing technology. As a result, Beijing government will be able to obtain a comprehensive view of urban operations, synergize applications and resources from different fields, and make intelligent use of information. The goal is to leap from digital Beijing to smart Beijing by intensive construction and application of the government cloud, thereby to reduce construction costs and energy consumption, and to improve the overall quality of operation and management of various types of e-government systems.

### 3.2.1.1 Contents of the construction of Beijing's government cloud

The IaaS platform of Beijing's government cloud electronic platform can provide integrated solutions for server, storage, switching, security, firewall, cloud platform, operation and maintenance, and so force [3, 4]. The intensive construction can effectively reduce cost and improve efficiency. Moreover, it is also green and reliable. The government cloud is based on a unified IaaS platform and allows for deploying the existing mature PaaS platforms, in order to further enrich government information applications. The government cloud also speeds up e-government construction by providing different users with different levels of SaaS applications.

- Using Liuliqiao data center and Beijing north data center (Miyun center) as a double design, where Liuliqiao data center is the main data center and analyzing center, and Miyun data center is the backup and archiving data center to combat disasters.
- Through rational planning of data center's network, we can transform operation and maintenance management from centralized into decentralized way with independent domains. Users of e-government are encouraged to manage and maintain their own websites through the

unified data center management and maintenance platform. In particular, the security design can help to keep a safe environment that can meet the security level in order to provide basic environment support for government applications and other security requirements.

• Managing and scheduling all the computing, storage and network resources by the unified management platform in a distributed cloud computing center can provide government departments with infrastructure services. It can also monitor real-time resource use, and make comprehensive analysis, rapid deployment and dynamic expansions. In this way, we can use resources in an efficient way and reduce energy consumption.

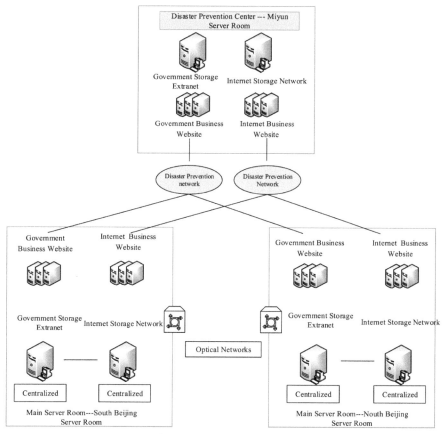

**Figure 3.1** Optical network.

- Designing a unified disaster recovery service to provide multiple levels of applications including disaster recovery and data backup services.
- Unifying the controls and the modules on the PaaS layer of government cloud platform to support the standardizing of common applications.
- Through the unified deployment of all types of common application components and service platforms, meet the individualized needs of various departments.
- Creating a directory of security services, including physical security, network security, computer security, virtual security and data security, application security and so on. Also included are institutional security, organizational security, personnel security, construction security, maintenance security, operational security and more.

### 3.2.1.2 Contents of the services of Beijing's government cloud

After the government cloud platform was launched, it started providing strong background protection for all types of government applications. By combining generic support components, the platform provides not only basic IaaS services but also PaaS services of a wide range and different levels. In addition, by cooperating with third parties, more products can be added to the directory of government cloud services to benefit users. Some examples of these integrated services are infrastructure, platform services, application services, value-added services, and security and management services. Bridging the isolated information islands between departments can allow users to enjoy truly intelligent, efficient and convenient services that are the new essence of smart city construction.

### 3.2.1.3 Functions of the application of Beijing's government cloud

Beijing government cloud platform is an administrative institution [5]. It not only provides a variety of functional IT services, but also connects government departments at all levels to safe and reliable IT services platforms. Other functions include promoting the intensive construction of e-government, eliminating information isolation and promoting information sharing, providing easy and convenient government services, enhancing the image of government, reducing financial expenditure, and facilitating the development of information industry. The government cloud platform provides cloud services to more than 20 municipal departments, including Municipal Organization

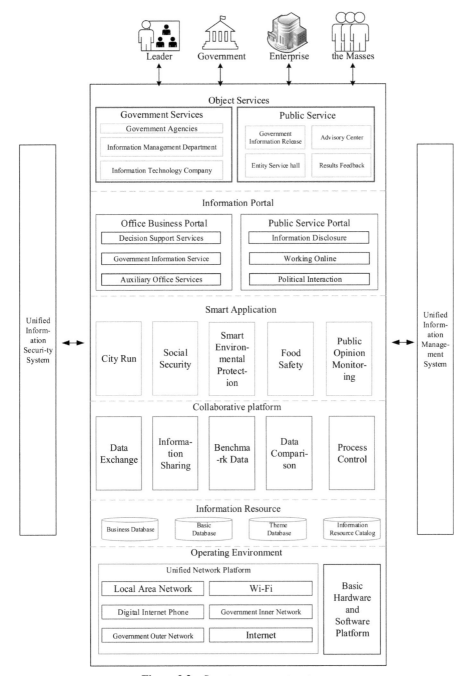

**Figure 3.2** Smart government systems.

Department, Municipal People's Congress Information Centre, Municipal Transportation Commission, Municipal State-owned Asset Supervision and Administration Commission (SASAC), and etc. More than 70 business systems have achieved secure and stable operations with the support of the government cloud platform. According to recent statistics, the platform has 99.99% accessibility and 99.9999% data reliability.

## 3.2.2 Case Study: Nanjing

Government data centre will complete the expansion projects of government internal and external networks so that the network bandwidth and access point bandwidth will reach 10 trillion and 1 trillion gigabits respectively, connecting

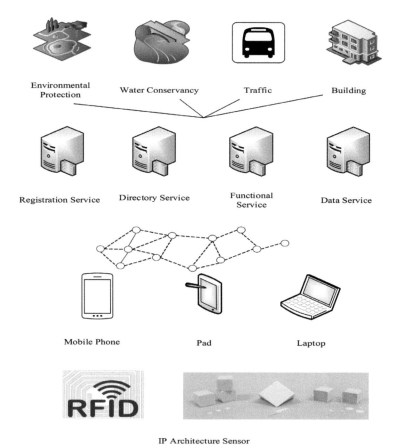

**Figure 3.3** The e-government.

221 municipal departments and 118 units, and reaching the computing power of 400 servers. Nanjing will also complete the construction of cloud computing service platform in order to provide support for the intensive municipal application systems. By self-building, self-collecting, and exchanging information with different departments, Nanjing will establish four databases of application information on demographics, corporates, government and urban operations. Particularly, the demographic information database brings together information of more than nine million people. The demographic information includes personal status, social insurance, health records, credit information and etc. The corporate information database has information of more than 200 thousand enterprises. The corporate information includes registration information, qualification information, and credit information. The government information database integrates information disclosure, transparent administration and performance appraisal information. Last but not least, the urban information database has pooled information on more than 20 departments. Examples of urban information include urban infrastructure, three-dimensional models, standard addresses, transportation, environment protection, water conservancy and so on. These databases greatly facilitate the management of smart cities and the application of e-government [6].

## 3.3 Smart Medical Care

With respect to medical issues, how to build an efficient medical and health system is a big problem. Although different countries have different ideas and methods of their medical and health systems, one common feature is that they have all adopted information technologies on their whole medical care systems and have digitized their health systems [7, 8]. The holistic goal of smart medical care is to provide residents with high quality health care services, continuous health information and comprehensive management. Smart medical care is composed of three parts, namely, smart hospital system, regional health system and family health system.

Smart hospital system consists of a digital hospital system with advanced applications. The digital hospital system serves all departments in the hospital. It collects, stores, processes, extracts and exchanges patient information and administrative information. The hospital information system, laboratory information management system and medical image storage and transmission system can collaborate in order to satisfy all authorized users. Smooth collaboration of the systems can provide the highest quality medical services, including rapid diffusion of disease information, easy transmission of basic information and remote assistance and so on.

The regional health system consists of regional health platform and public health system. The regional health platform can be accessed from mobile phones and computers, and therefore transfer information recorded by communities, hospitals, research institutions and supervision departments. It utilizes modern technologies to create electronic files of patients and properly manage these files. With the help of computer and network communication technologies, hospitals in different regions can share patient information and therefore ensure timely and accurate medical treatment and program for patients. In addition, the public health system functions as the regional health care supervision department. It can easily acquire the latest information on the health situation in the region, enabling timely response to any events and emergencies. In this way, the regional health system can deal with potential diseases and epidemics in advance and respond in time, thereby minimizing risks.

The family health system is closely related to public's daily health protection and is mainly for those without easy or convenient access to hospitals for treatment. The system enables remote care of aged people and remote data transmission monitoring. For those who cannot take care of themselves, health care workers can real-time monitor patient information via the automatic remote data transmission prompts.

With the development of medical technology, hospitals have accumulated a large number of medical data of different forms, including audio, video, images and etc. These data have become a valuable asset of hospitals. Medical data are commonly unstructured or semi-structured data in text, images video, and multimedia forms. According to estimates, a medium-sized city in China can accumulate 10 PB medical data in 50 years. As health care continuously develops, the growth of unstructured data will also accelerate. The traditional relational databases will no longer have advantages in storing, processing or searching in large datasets. It is urgent to design and upgrade the data storage platform, management and searching platforms for large datasets.

In Hong Kong, smart health care system was launched in 2007 [9]. The core is promoting electronic health records and creating easy access to them. Electronic health records, also known as electronic medical records, rely on a system that saves information related to personal health status in electronic form. The system includes not only medical records but also personal health records. The benefits of implementing electronic health records are obvious. For patients, electronic health records can provide the complete record for doctors to make comprehensive diagnosis, thereby reducing duplications of inspections and treatments. Electronic records can reduce total medical cost so as to improve efficiency. For doctors, they can acquire comprehensive patient information and thus reduce the risks and errors of handwritten medical records. Finally, for public health departments, electronic medical records can help acquire fast and timely information on public health and safety situation, and make fast and accurate response to public health emergencies.

Despite that public hospitals in Hong Kong have already been practicing electronic medical records and easy access of patient information, the government of Hong Kong still attempts to further increase the accessibility of electronic medical records with big data technologies so as to truly serve the people in Hong Kong. In order to fully employ the accessibility of electronic health records (EHR), the Hong Kong Hospital Administration Authority has launched a public-private partnership (PPP) program that covers both electronic medical records and health records. The program will enable registered private health care providers to access patients' electronic health records on a secured platform with the consent of patients. It has provided an electronic platform for private and public hospitals to collaborate. By the end of 2013, many private health care providers have joined the program. Electronic health records clearly are an important part of the program.

It can be said that Hong Kong has embarked on a fast track in popularizing electronic medical records. We believe that in the near future, Hong Kong will be able to achieve the goal of personal lifelong electronic medical record, which is an ideal situation of modern health services.

### 3.3.1 Case Study: Hangzhou

Since 2012, more than 2.5 million residents in Hangzhou have started using the electronic billing function of their city card. Accumulatively, smart medical care has benefited 6.27 million hospital visits. Assume that each hospital visit can save the labour activity of all persons involved with one hour, then the smart medical care program in Hangzhou reduces the time costs with more than 6.27 million hours. The online health examination and result checking systems have served 953 thousand people in total. At least 10% of the population are free from travelling back and forth to hospitals for examination results [10].

Since Hangzhou launched its smart medical care program, patients can make online appointments through hospital information systems, and can choose the exact time for their visits. As a result, patients can choose the doctors they want to consult for advice and treatment according to their own

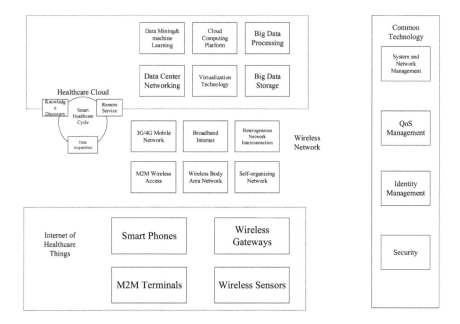

**Figure 3.4**    An overview of smart healthcare in Hangzhou.

time schedules. Departments of ultrasound, CT, MRI, endoscopy, and more. are also available for online appointments. In this way, patients can skip the long waiting time in the hospitals and enhance their hospital experience.

In 2014, Hangzhou Red Cross Hospital has piloted in implementing the one-stop booking centre. The booking centre accept reservations of various medical services and can make appointments for patients at their convenience. The well-trained booking centre personnel will also remind patients of relevant precautions, and therefore bring great convenience to patients.

Meanwhile, Hangzhou government has collaborated with smart cards manufactures and launched multi-functional self-service automatic machines with unified logo, unified appearance, and standard functions. These self-service ATMs are easy to use. People can apply for city cards, for services such as to make hospital appointments, register, search, pay bills, print lab results and do many other things via the machine. Recent statistics show that the average utilization rate of these hospital ATMs has reached 76 percent. Number 2 Municipal Hospital receives the highest utilization rate of 94.72 percent. On average, 68 percent of the hospital appointments are realized via ATMs.

At the moment all municipal hospitals have adopted self-service areas. Some hospitals even closed the registration window in the reception hall, only keeping one or two traditional manual service windows. Patients of the municipal hospitals have gradually abandoned the traditional manual service windows, and become more and more willing to choose for self-service.

### 3.3.2 Case Study: Real-Time Cardiac Remote Monitoring Platform

The remote monitoring platform can automatically collect data of various vital signs, upload these data to the hospital control centre, make real-time data analysis and early warnings, and facilitate doctors to provide remote medical services. Using different portable devices, data can be collected without time and place restrictions.

The remote monitoring system can monitor and detect symptoms of diseases in cardiac functions, urination, blood pressure, blood sugar, sleep and so on. In this way we will introduce the real-time cardiac monitoring system and sleep monitoring system [11].

The World Health Organization (WHO) has pointed that cardiovascular diseases (CVDs) account for nearly 30% of deaths, and is the leading cause of death worldwide. It is estimated that in 2008 alone, 17.3 million people died from CVDs. More than 80% of CVD deaths occur in low- and middle-income countries due to inadequate preventive measures and lack of access to health services. By 2030, 23.6 million people will die from CVDs each year. In China, the number of CVD patient has reached a hundred million, accounting for about half of all deaths.

According to statistics from Beijing Emergency Centre, more than 70% of CVDs occur at patients' home or workplace. Due to the lack of timely aid, such morbidity is very likely to lead to death. Therefore, it is necessary to change the traditional method that prioritizes treatment, and emphasize prevention instead. That is to utilize various medical IoT technologies to provide continuous monitoring and early warning for the CVD high-risk groups.

At present, hospitals commonly use static or dynamic electrocardiogram (ECG) instrument to record and compile the ECG changes in active and quiet states. However, these traditional instrument and equipment are difficult to carry with the patient and may fail in long-time data collection. Moreover, data can only be stored in the instrument until collection is completed and

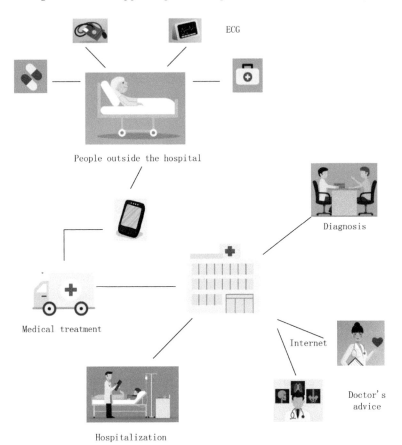

**Figure 3.5** Remote monitoring platform.

the data is ready for analysis. As a result, patients cannot receive real-time feedbacks on their cardiac functions.

In order to solve these problems, we have designed a real-time cardiac monitoring consisting of three components: ECG collector, mobile phone and medical service centre. The ECG collector is a portable device with high accuracy, small size, low power consumption, and high automated level. This device can measure and record heart rate and other vital signs, and transmit the data first to a mobile phone via Bluetooth, and then from the mobile phone to the medical service centre through the wireless network. The software at the medical service centre will identify abnormal status and immediately report to doctors and patients. Finally, doctors can provide patients with remote services, such as detecting risks and giving advice via mobile phones. In addition, some mobile devices have advanced computing power. They can analyse real-time ECG and properly store and manage historical data. It is worth pointing out that we have designed automated analysing software that can detect more than 20 kinds of abnormal ECG patterns and adjust the parameters according to specific circumstances, to provide patients with personalized service.

## 3.4 Smart Communities

Communities are the most basic component of cities. They are also the bases where urban residents can survive and develop. Informatization within communities is crucial for urban intelligence. From the functional point of view, residents in the community are the core of the community service. Communities should be able to provide residents with safe, efficient and convenient intelligent services. Smart community is composed of a highly developed service centre, high-level security, and intelligent community management [12].

Smart community construction utilizes various technologies including internet technology, financial technology, cloud computing technology, big data processing technology, and data mining technology, to integrate network communications, smart appliances, home security, property services, community services, and many other value-added services in an efficient system, so as to create safe, comfortable, convenient, environmentally friendly, intelligent and humane living environment for community residents, and to provide comprehensive information services to them.

To facilitate the development and management, we divide smart communities into three layers from top to bottom, namely, the application layer (Software as a Service, SaaS), calculation, integration and exchange layer (Platform as a Service, PaaS), and platform support layer (Infrastructure as a Service, IaaS). Particularly,

- The SaaS layer provides applications for smart communities. For example, create a number of portals to smart communities, provide community management and services and more, and enable platform users to subscribe to the portals and applications.
- The PaaS layer supports open component and third-party applications. For example, provide SaaS layer with functional component services such as Business Intelligence (BI) support and application engine,

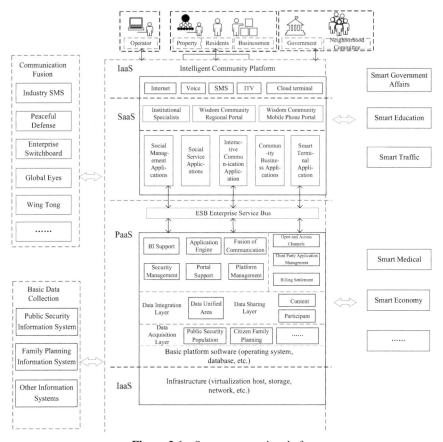

**Figure 3.6** Smart community platform.

integrate the internal and external applications of the platform, and provide data application and data sharing. This layer contains basic software such as operating systems, databases and middleware.

- The IaaS layer provides access to residents, properties and communities, and is constructed on a cloud platform where there are various infrastructure clouds.

## 3.4.1 Case Study: Smart Communities and Smart Families in Shanghai

In June 2011, Shanghai invested 30 million Yuan to build the first smart community Pudong Jinqiao Biyun, where in the first period the transformation has been completed with four major achievements: smart home terminal, Jinqiao Biyun card, community information portal, and the cloud computing centre. The smart home terminal "Biyun housekeeper" has many functions such as public service information searching, discount information prompting, and service booking and so on [13]. People can directly pay bills, make reservations and enjoy personalized services by subscribing to various business and community service providers with their Jinqiao Biyun cards. The community information portal is the online window for residents to receive on all kinds of information in the community. This function is similar to Biyun housekeeper. Meanwhile, thanks to the interactivity and communication of websites, these services can be expanded to serve all groups. The cloud computing centre is the brain of the entire project, because all the sub-projects need to exchange, process, store and search data in the cloud computing. In addition, the program also contains a smart traffic project where traffic lights are monitored to detect violations of traffic rules, a smart environment project where relevant departments will receive notice when the community trash bins reach 90 percent usage, and a smart parking project which can provide quick information on parking lots and parking guidance.

In Chinese culture, family is the smallest social unit. At present, Shanghai Telecom is focused on promoting smart families in the city. The most fundamental practice is to speed up home broadband internet. Turn on the Internet Protocol Television (IPTV) at home, people can complete community hospital registration via the digital IPTV terminal. In case of emergencies such as earthquake and fire, the community alarm system will immediately send signals to residents' phone or mobile terminals. The home terminal also enables people to share their travel experience with families and friends at any time through the mobile network. Last but not least, it is also possible to monitor parking spaces and make it easier to find parking spots.

Since 2013, the Shanghai smart community project will be extended to more than 2000 communities in town and benefit millions of residents, so that they will be able to enjoy convenience in all basic aspects of daily life. This is an example to show how smart communities are the initial step toward smart cities. The smart city that Shanghai Telecom committed to build is gradually taking shape.

### 3.4.2 Case Study: Taiyuan County, Taiwan

Another example of smart communities is Taoyuan County in Taiwan. Taoyuan is located in the northwest of Taiwan. In the past, peach trees were planted all over the county. When peach blossomed, the area became colourful and fragranced, and therefore was named Taoyuan (meaning the peach stream) in 1886 (12th year of Guangxu, Qing Dynasty) officially [14]. In 1941, Taoyuan became a county. The landscape in the county can be divided into coastal plains, hilly terraces, and mountain terrains. The county has an area of 1220 km$^2$ and a population of 2.04 million (2013).

In 2013, the Intelligent Community Forum (ICF) of the New York think tank selected 24 cities out of 400 globally to examine the detailed socio-economic development of communities in the 21st century. Seven cities were shortlisted, and Taoyuan ranked number six.

The smart community project in Taoyuan is also known as the M-Taoyuan project. In order to meet residents' information need and to provide more convenient information exchange method, Taoyuan administrative authorities have been actively designing the M-Taoyuan project since 2005. The project is composed of four sections and 17 wireless broadband-related application services. The four sections are facilitating daily life by mobile broadband, promoting industrial and commercial marketing to enhance performance, real-time monitoring and reporting of security status, and mobile working. It is noted that the 17 services include a wireless network infrastructure.

Among them, the mobile broadband infrastructure is based on the public facilities at the county's administrative authorities. A total of 30 WiMAX base stations are built, as well as a T-shaped Worldwide Interoperability for Microwave Access (WiMAX) mobile corridor of 37.5 km$^2$ that covers the core metropolitan area of Taoyuan. The rest of the region integrates WiMAX, Wi-Fi, 3G and other internet services, so that residents who live outside the WiMAX coverage area can also use M-based application services. The administrative authorities of Taoyuan County have been actively building Wi-Fi hotspots, WiMAX mobile corridors and 3G-related environmental facilities. The main purposes are to provide a variety of network interface to residents

**Figure 3.7**   "i-Taoyuan" project.

in the county and to establish convenient information communication channels. The administrative authorities hope that through the construction of WiMAX mobile corridors, they can accelerate the integration of wireless broadband technologies in Taoyuan County, in order to attract relevant industries to settle in Taoyuan.

In terms of promoting convenient life, the county authorities have been building Mobile Instant Pages (MIP) mobile navigation site to allow residents quickly access six sections of services, i.e., M-public services, M-broadband life, M-business applications, M-public security bulletin, M-sightseeing service and Tahiti project with their mobile devices. Taoyuan residents will be able to have convenient access to various types of local information via mobile phones or tablets (PAD).

The administrative authorities in Taoyuan have adopted innovative applications of mobile technology on fire-fighting to facilitate the fire control personnel to obtain fastest information of buildings regarding their fire configuration. The mobile fire safety inspection control system is one example. Its action command system can return disaster and order on-site images and videos immediately via mobile devices to facilitate the remote control and decision-making of the site. In addition, the information feedback system and rapid notification system of public sectors can further increase government efficiency.

Industrial and commercial applications are mainly used to strengthen online marketing channels, enterprise mobile learning environment and remote video connections.

In order to better implement the vision of Smart Taoyuan, in 2009, the county has started the "i-Taoyuan" project [15]. The i-Taoyuan project focuses on Taoyuan Aviation City, which integrates passenger transportation, freight transportation, production, sightseeing and living to promote the construction of high-quality smart city demonstrations. The Taoyuan Aviation City has a total area of 6150 km$^2$ and is expected to receive 2.4 trillion Taiwan dollars investment. Eight areas have been planned, including the airport area, the aviation industry area, the free trade port area, the economic and trade exhibition area, the airport compatible industrial park, the coastal recreation area, the fine agricultural development area and the living area. The authorities of Taoyuan county hope to build an intelligent and forward-looking aviation city that can fully integrate digital infrastructure and supporting services through infrastructure construction. The proposed infrastructure includes regional digital optical fibres, wireless broadband networks, intelligent transportation,

**Table 3.1**   Network systems in Taoyuan

| Application | Content |
|---|---|
| Smart business | Online customer service, electronic convenient stores, county information searching and reporting system |
| Smart transportation | Smart transportation management system, smart buses |
| Smart education | Smart teachers, cloud e-learning |
| Smart medical care | Smart medical care passport |
| Smart life | All-captured video equipment, GPS dispatching system |
| Smart sightseeing | Audio and video travel information |
| Smart environment | Air pollution detection station, water quality testing network, smart building expert group |

security monitoring systems, and 4G mobile broadband networks that will integrate WiMAX and LTE. Based on the smart Aviation City, the administrative authorities in Taoyuan have been continuously expanding the scope of communication technology applications to gradually form an information network system covering all aspects of life, such as medicine, food, housing and transportation (Table 3.1), and providing economic, convenient, and safe services to all residents.

From the above case we can see that the primary element of construction smart communities is building a basic network which can provide community residents and visitors with the most convenient internet access. Based on network construction, we can launch all types of smart applications. The goal of smart community construction is to provide residents with safe, efficient and convenient living environment. As the number and types of smart applications increase, we can predicte that massive data will be generated on smart community platforms. Using big data technologies to make in-depth excavation and analysis of these data will be an important development theme of smart community construction in the future.

## 3.5 Smart Transportation

Intelligent Transportation System (ITS) is a concept brought up by the United States in the early 1990s [16]. It effectively integrates advanced information technology, data communication transmission technology, electronic sensing technology, control technology and computer technology and applies to ground traffic management system, so as to establish a real-time, accurate and efficient comprehensive traffic management system of a wide scope and various functions.

Smart transportation is a concept proposed by IBM in 2009 [17]. Based on ITS, smart transportation integrates IoT, cloud computing, big data, mobile internet and other new technologies to collect traffic data and provide real-time traffic information services.

With the acceleration of urbanization, urban traffic faces tremendous pressure. Shortage of land resources is restricting the sustainable development of modern transportation. Energy consumption caused by overwhelming "automobile-ization" and the lag of public transport development lead to contradiction between transportation supply and demand. Since 2012, the annual economic losses caused by traffic congestion have reached trillion scale. Traffic accidents are frequent, and traffic pollution is becoming more and more serious. Applying big data, IoT, cloud computing, mobile internet and other new technologies to the transport sector will greatly enhance the operational efficiency of the transport system given the existing transport network. In Japan, for example, the intelligent transportation systems that involve new technologies can increase the capacity of existing transportation network by about 20 percent, while in the United States, the number is 16 to 62 percent.

Most urban traffic management in China is segmented. Because traffic data only stores in the so-called "vertical businesses" and in single application, the result is the fragmentation of traffic management. Common problems include traffic information dispersion and too simple information content. Big data technologies have advantages in information integration and scale efficiency, and therefore are helpful to synchronous data acquisition and processing. With the introduction of intelligent traffic sensors, the scale of data has exploded to PB level from the past TB level, bringing new challenges to the storage and calculation of massive data. It is urgent to find new processing technologies and methods. At the same time, in the transportation sector, demands for real-time processing and analysing massive image data and video data are high. Combined with IoT and cloud computing technologies, big data analysis plays an important role in the intelligent transportation system. We will see some typical example applications of big data analysis in intelligent traffic systems.

### 3.5.1 Case Study: UPS Vehicle Control System

One example of big data application in smart transportation is United Parcel Service (UPS)'s vehicle control system [18]. In order to enable its headquarter to trace and locate vehicles, and to prevent engine failure when there are delays, all UPS trucks are equipped with sensors, wireless adapters and GPS.

This is also one way to facilitate company's supervision and management of employees, as well as to optimize routes. The optimal routes customized for trucks are to some extent based on past traffic statistics. With the help of vehicle control systems, UPS drivers accumulatively drive nearly 48.28 million less km in 2011, saving 3 million gallons (1 gallon = 3.7854 litres) of fuel and reducing carbon dioxide emissions by 30,000 tons. In addition, the smart system is designed to make as little as possible left-turns because left-turns at the intersections require trucks to cross the road and therefore will likely lead to more traffic accidents. Moreover, trucks often need to wait for some time to turn left, which will consume more fuel. Therefore, reducing the number of left-turns can greatly improve driving safety and efficiency.

### 3.5.2 Case Study: Smart Bus Transportation of Beijing

In the "Internet+" era, it is necessary to improve the operations of transportation hubs with intelligence-oriented thoughts, so that they can provide stable support for cities. In this context, smart transportation system is invented.

The solution to urban traffic problem is building and dispersing. Promoting smart transportation will be the best way to improve the transportation efficiency, to alleviate traffic congestion, and to reduce traffic accidents. From the experience of worldwide practice, the development of smart transportation system indeed demonstrates the mentioned improvements and at the same time protects urban environment and conserve energy.

In order to ensure the security and stability of Beijing's transportation system and improve the capacity, Beijing launched the first Transportation Operations Coordination Center (TOCC) in China [19]. The centre integrates dynamic traffic monitoring, video resource management and application, and unified public information. In the centralized command hall, with Geographic Information System (GIS) technology, cloud computing, GPS and so on. We can real-time monitor, analyse, manage and publish comprehensive information of traffic operations and flow data of the city. At present, TOCC has access to more than 6000 kinds of traffic data, covering urban road network, bus, taxi, subway, airlines and other fields. The centre has therefore achieved the goal of smart management and monitoring of urban transportation. By continuous strengthening its smart transportation construction, Beijing has greatly improved its transportation services and the degree of intelligence.

## 3.5.2.1 Safety

- Buses: more convenient, more accurate, and more comfortable experience

In the recently released Social Responsibility Report 2015 by Beijing Public Transportation Group, the "Internet+" model will become a new topic of urban public transportation in the future.

By the end of this year, more than 6000 buses will install simple alarm systems. In case of emergencies, the alarm systems will send live videos to the public security department. At the same time, some pilot buses will adopt the Internet+Bus mode and increase their carrying capacity, to improve punctuality and comfort of buses.

- "One-click" simple alarm systems on 6000 buses

When unexpected situations happen during driving, bus drivers will only need to press a single button to connect to public security department and send videos and audios to the public transportation emergency department for rapid response. This year, Beijing will install this one-click simple alarm systems on more than 6000 buses. Buses that are installed with the system will be monitored when they are in use. Meanwhile, the public transportation department will also build an emergency image and video management system to improve the ability to handle emergencies.

- More accurate and comfortable

In 2016, the so-called public transportation e-pass mobile APP has covered all bus lines. People can real-time check departure and arrival information of all bus lines on mobile phones, and then plan their trips to avoid waiting for long time. In the near future, the Beijing Public Transportation Group will promote "Internet+ bus" convenient transportation, upgrade the public transportation e-pass mobile APP, improve network efficiency, and better serve passengers.

At the moment the Beijing Public Transportation Group is working with Tsinghua University on related topics to explore the further development of public transportation in the context of Internet+. They also aim to improve the public transportation e-pass mobile APP and also their customized information platform with the help of big data technologies. Meanwhile, it is possible to predict passenger flow and adjust the network according to daily public transportation statistics. This means to allocate carrying capacity efficiently according to actual volume of passengers and operating conditions through data collection and analysis. Even in the case of traffic jams, the punctuation

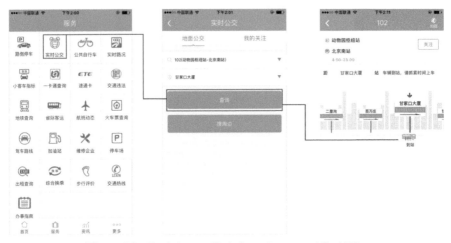

**Figure 3.8**　Real-time traffic information on mobile APPs.

of buses can be guaranteed. By then, public will be able to check bus arrivals and departures on their mobile devices and calculate the time to go out. In this way, they will get on the bus on time. The vision of the Beijing Public Transport Group is: "We will be able to analyse the most concentrated time slots and stops as long as passengers use the electronic bus card. Then we can make full use of these data and make proper allocation of operating capacity. Subject to the actual traffic conditions in Beijing, in the future buses may not be faster, but will be more punctuate and more comfortable". This year they will be promoting the "Internet+ bus" model in some pilot bus lines.

- Big data analysis is applied to public transportation

"We cannot change the road congestions, but can adjust the bus operations according to the actual road conditions," according to the Beijing Public Transport Group. Since the end of 2014, Beijing has started level-pricing of bus fares. By analysing bus card data, the system can determine the public transportation utilization rate at different periods and different regions. Big data analysis can also tell us which bus lines have sufficient capacity and which ones are relatively idle.

Based on these data, the dispatching system will complete an upgrade. The originally fixed bus terminals will become more flexible. Specifically, the public transportation department will be able to schedule more buses at any time according to traffic and passenger conditions, in order to reduce passengers' waiting time.

Last year, Beijing has removed repeated bus routes of 223.26 km on the Third Ring, and thereby reduced the traffic pressure of the Third Ring. This year, the extension of bus lanes on the Third Ring will further improve the convenience of travelling with high-capacity buses. We believe that this will attract more people to travel green.

- Internet+ public transportation

The Beijing Public Transport Group focuses on improving the operational efficiency of bus lines. Currently the group is working on several projects, including creating a non-stop route to connect Beijing, Tianjin and Hebei, adding two new tourist routes, improving the real-time bus information searching function and the customized platform, and launching bus information on WeChat. In short, the group is making efforts to provide passengers with diversified, accurate, and customized bus services.

### 3.5.3 Convenience

Recently, Beijing Municipal Transportation Commission launched its official mobile application Traffic Beijing. In the first phase the application has 20 functions including real-time bus information, public bicycle availability, real-time traffic, minibus index and so on. These functions cover bus, subway, public bikes, car, taxi, railway, airlines, and provincial buses. The mobile application can sufficiently serve the needs of public transportation.

Real-time software of the bus schedule has been commented as not reliable. Renewal of this mobile application enables an updated version of real-time bus schedule query system that refreshes bus locations every 5 seconds. As a result, the forecast accuracy has been increased by 10 percent and reaches 85 percent now.

At present, live information of 613 bus lines can be accessible through the mobile application. 611 lines are ordinary bus lines, and the other two are airport lines. Wangjing Line and Sihui line are also integrated into the real-time system for the first time. By the end of the year, more than 900 bus lines in Beijing will become real-time accessible.

The Municipal Transportation Commission promised that it will continue updating the mobile application, and it is expected to add new services and functions such as customized buses, railway traffic congestion, and inquiry service on traffic administrative licensing.

The metro congestion index function is expected to be achieved soon. And during this year it is expected to realize customized buses function.

At present, Beijing railway transit network has been well constructed. By choosing different transfer stations, passengers from A to B can have different route options. If they can check the congestion index of each subway line, they can avoid wasting time. The public bike availability function can give information about the distribution of public bicycles in the city, the available number and their specific locations. People can also use the review function to recommend the best walking route.

In terms of real-time traffic, the Municipal Transportation Commission and the Municipal Traffic Management Bureau have merged their data for the first time to create a road traffic information map with higher accuracy. 93 percent of roads within the Sixth Ring were covered by real-time traffic information system now, compared with 87 percent in the past.

### 3.5.4 Free Wi-Fi on Buses

From now on, buses in Beijing will have access to free Wi-Fi. Bus Wi-Fi operator 16WiFi recently announced that bus Wi-Fi hotspots have obtained public security certification and network security certification. Currently, Beijing's No. 300 bus line has completed equipment upgrading and is officially providing free Wi-Fi on buses. It is expected that by the end of September 2017, 18,000 buses in Beijing will all have Wi-Fi access. Passengers can enjoy free Wi-Fi on buses. For example, they can chat and shop online, and even watch video.

In fact, this is the second attempt to provide free public Wi-Fi in Beijing. In 2013, One Way Wi-Fi, the predecessor of16WiFi, tried to widely set up hotspots in Beijing buses, but did not succeed due to immature technology and poor experience. Qiu Zhaomin, the founder of 16WiFi, said that 16WiFi has upgraded its equipment and business model in 2016 to ensure that the situation in 2013 would not happen again.

The new attempt has solved previous problems, said Qiu. In terms of hardware, the upgraded car routing devices can distribute 3M bandwidth to each connected user to enable an actual downloading speed of up to 400 k/s. This is enough for 50 people to access internet smoothly at the same time. Browsing websites, chatting, and e-shopping have become easy. People can even watch HD videos at the same time. In addition, registration is quite simple. Users only need to download the 16 WiFi APP, register, and then connect to the bus hotspots (Figure 3.9).

It is expected that by the end of September, over 80 percent of all bus lines in Beijing will have free internet access. In the future, the 16 WiFi APP for

**Figure 3.9**   Instruction of accessing free Wi-Fi on buses.

free bus Wi-Fi connection will add an arrival reminder function to prevent concentrated passengers from missing their stops.

### 3.5.5 Case Study: Internet+ Operation of Taxi

Didi Chuxing is one way to connect taxi drivers and passengers with smart phones [20, 18]. Based on this platform, passengers and drivers can mutually evaluate, in order to establish a credit system. The system can facilitate rapid information exchange, improve passengers' travelling efficiency and drivers' operating efficiency, and provide a good solution to some of the problems and weaknesses in the taxi industry. The taxi service of Didi is available in more than 300 cities in China, with over one million registered taxi drivers in total, and more than 100 million registered passengers. The peak number of orders in a single day was 5.3 million. Didi Chuxing has reduced on average 20 percent no-load rate of taxi, increased the income of 94.7 percent of the drivers with 10 to 30 percent and the rest time by an average of 100 minutes. Moreover, people who use Didi Chuxing have increased their chances of hiring a taxi successfully to 85.8 percent with a 5.4-minute average waiting time only. Nationwide, the 1.3 million taxis have reduced no-load driving distance by 40 to 50 km per day, saving 4 to 5 litres of fuel and emitting 7.3 million less tons of carbon annually, which is equivalent to 600 million trees in annual ecological compensation.

On September 1, 2016, Didi Chuxing officially released its traffic cloud computing platform. Zhang Wensong, the senior vice president of Didi said that the platform will provide smart transportation services to support urban transportation construction and public's travel decisions. According to the transportation big data collected on the smart traffic cloud platform, they can enable many functions such as regional thermal maps, OD data analysis, urban transport capacity analysis, urban traffic forecasts, urban traffic reports and dynamic timing of traffic lights. The platform can also help to improve public transportation by providing information about real-time traffic, real-time bus arrivals, Estimated Time of Arrival (ETA) and urban operation capacity. Zhang Wensong further introduced that by mining of massive data of drivers and passengers, combined with advanced artificial intelligence algorithms, Didi can real-time calculate the best route to reduce travel time and travel costs. At the moment, Didi can make dynamical predictions of traffic demand for the next 15 minutes, leading drivers to approach those high-demand areas in advance. In this way, Didi can increase the number of successful orders per hour and enhance travel experience.

Similarly, the high efficiency operating capacity supported by big data not only ensures that every single demand is met to the maximum extent, but also contributes to the functioning of carpooling system. Based on the massive data on Didi's platform, we can set up and optimize the virtual sites to improve the quality of carpool services and to achieve high efficiency.

The application of big data on Didi's taxi service is a pioneer innovation, especially in improving the service quality of taxi drivers. In September 2016, Didi launched its service credit system in 100 cities nationwide. This system is an example of using big data to identify drivers who provide outstanding services, and help them generate more income. Zhang Wensong pointed out that the matrix service sub-system has remarkably increased the percentage of high-quality drivers. Higher service score helps drivers to receive higher income.

In addition to improving the efficiency of taxis, Didi's smart traffic cloud has also been promoting the implementation of Internet+ transportation actively. Over the past year of 2016, Didi has been collaborating with a number of local governments to build transportation big data platforms to merge government and private data smoothly. Currently, Didi has assisted Jinan and Guiyang governments to establish real-time traffic data system. It is said that in future, the two parties will extend collaboration on the smart traffic cloud to achieve road network optimization, intelligent control of traffic lights and so on. This will add much value to the public transportation system of the cities.

In the future, Didi's smart traffic cloud will integrate its Origin Destination (OD) data, driver data, GPS trajectory data, and operating capacity data, as well as sensor data, static road data, and road events data, to provide better services for cities.

## 3.6 Smart Security

Whether a city can provide residents with life and property security is an important factor nowadays when people choose which city to live. The level of urban crime rate, whether there is an effective anti-terrorism mechanism and whether there are effective measures to deal with emergencies are important considerations of urban security. At present, in the event of major public security issues, the common response in China is to hold emergency meetings chaired by one or several senior government officials and form a temporary command group to deal with public security emergencies after events. The response on emergencies always has a time lag and the way to respond may be too subjective without scientific analysis of information. Under the current security situations, government must make efforts to maintain and promote urban public safety more actively. Specifically, to reach higher level of urban security, the following aspects must be addressed. First of all, we must aim scientific development of the city security. Secondly, in the design phase of new cities, we need to be guided by specific public security policies and make rational and scientific design. We also need to establish a mechanism for long-term, effective and safe information communications to ensure the effectiveness, promptness, and clarity of information transmitted. Last but not least, we should create an incentive mechanism to encourage residents to participate actively and to avoid moral deficiencies.

To ensure security, the following practices need to be taken into account:

- Installing cameras in main streets, installing monitoring equipment in dangerous zones and increasing the number of monitoring points and perception systems.
- In the construction phase of new infrastructure, the mobility of traffic should receive sufficient attention to ensure that maintenance personnel can arrive in a timely manner in the event of emergencies.

- Ensuring timely and effective information sharing and secured transmission, in order to facilitate coordination from the upper level. Taking specific arrangements to minimize losses.
- Instead of relying solely on governments, involving more urban residents and government agencies to jointly improve the response to public safety issues.

The construction of urban security cannot be separated from information science. Our information age has added a new connotation to the management mechanism of urban public security. Protecting the lives and properties of residents requires collaboration between governments and information technologies such as sensors, RFID, network communication and video detection, big data analysis, and so on. In this way, we can achieve security analysis, early warning, security management, and security protection. For example, the public security departments have collected a large number of video image data via the Skynet system adopted nationwide. Using these data, they can track the whereabouts of suspects, such as where in the country they have been and at what time. Relying solely on human power is not enough to fulfil such tasks due to the limitations of human power. Some of the key technologies and systems involved are face recognition and image recognition, model processing, data compression, and massive data processing. They help police officers to extract relevant information and to improve their work efficiency.

One of the most important features of smart security is the detection of potential threats. The focus of "safe city construction" has changed from ex-post investigation to ex-ante early warning. In the era of big data, data sources can be tremendous. The proportion of unstructured data such as voice, video and image increases gradually. Massive data are closely related to our lives. Internet has carefully recorded human behaviour, locations, consumer information and so on, so that it captures nearly all details of behaviour and privacy. Massive video monitoring data are the core of smart security. Precise analysis of these data is the key function of big data application in safe cities.

Safe city data has its industry-specific requirements. First, it is a huge flow of information. Second, it requires high resolution. A high-definition video is generally 8 MB. Safe city data come from thousands of cameras and are 24-hour continuously transmitted. If we use a centralized computing framework, data transmission blocks are likely to happen. In extreme cases these transmission blocks can lead to catastrophic losses. On the other hand, fully a decentralized computing framework will lead to complexity in management, operation and maintenance. Therefore, regional and integrated centralization, and conceptual decentralization will support network applications in a more effective way.

The rapid growth of big data requires corresponding growth of IT infrastructure in terms of computing power, storage capacity, data exchange and transmission capacity in order to ensure data analysis. The key to achieve the growth is the scalability of IT infrastructure, particularly, the ability to scale horizontally. Powerful horizontal scalability can lead to advantages in cost and performance.

Currently, all countries are attaching great importance to big data analysis in smart security applications. For example, data analysis is effectively applied to disaster prevention in Korea. Korea's permanent disaster prevention agency, National Emergency Management Agency (NEMA), has been headed by a central committee where the Prime Minister of Korea is the chairman, the ministers of finance and economic affairs and the minister of security administration are deputy directors since its founding. The agency is responsible for disaster prevention related work. In the event of disasters, it will collect meteorological information, hydrological information and other disaster information nationwide, and provide timely rescue and prevention through big data analysis. Such integrated data analysis is based on aggregated data from various departments of the Korean government. At the same time, the release of disaster information in Korea also start paying attention to new technologies and the numerous mobile phone base stations, since the mobile phone signal basically has no blind spot. Last but not least, the agency can obtain the latest information of the disaster in an effective way simply by collecting and analysing the signal data.

In Moscow, Russia, the urban railway system extends in all directions of the city, on and under the ground, thereby constituting a highly developed transportation network. Residents in Moscow normally choose to travel by car, but when they encounter bad weather, the security of the urban transportation system will be under great pressure. Moscow Meteorological Bureau is responsible for warning all residents of potential dangers. They can achieve this by analysing the disaster data with modern high-tech transmission methods and information processing technologies. For example, when the bureau forecasts or monitors heavy snow and other emergencies, it will immediately inform relevant municipal government agencies and other units. It also provides disaster information to various medias.

In Taipei City, smart security has been regarded as an important element in smart city construction. Taipei has established an information integration centre to process all relevant information on urban security. The centre has adopted an integrated security analysis system and a mobile security monitoring system. The security monitoring system includes M-Police, instant video transmission system, and a 16 million pixel mobile high definition video monitoring system. These devices are playing a significant role in maintaining urban security. Through the large amount of information collected by the integrated security analysis system, it is possible to identify potential security hazards in advance in order to ensure the security of Taipei.

Adopting big data technology not only can signal early warnings of natural disasters and security emergencies, but also can provide full protection for people's daily life. Below we give an example of private security which involves available technology and big data analysis. In Japan, Professor Kosei Shimizu from the Institute of Advanced Industrial Technology works on studying people's sitting position. Few people may think that a person's sitting position can contain any information, but it really can; also they may think has it any relation to private security and it really has. When a person sits, his stature, posture and weight distribution can be quantified to generate data. The professor and his team of engineers installed a total of 360 pressure sensors in the lower part of car seats and measured the pressure on the chair. They successfully converted human body characteristics into valuable data and quantified by a range of values varying from 0 to 256, so that each subject had his or her personal data exclusively. In this experiment, the system can recognize differences in seat pressure and identity different passenger based on the pressure data with an accuracy rate of up to 98%. Similar technology can be utilized and installed on cars as anti-theft system. With this system, cars will be able to identify whether the driver is the owner or not. If not, the system will ask the driver to enter the password. If the driver cannot enter the password correctly, then the car will not turn on. Once we have extracted data from a person's sitting position and many other aspects, these data can inspire and create a number of practical services and even an industry with bright future. This will also serve as a very important practice of ensuring the security of properties, as well as the security of cities to a certain extent.

### 3.6.1 Planning and Construction Status of Chinese Smart Cities

China's smart city construction began in 1995, when geographic information system was the core of the construction [21]. It can be called the 1.0 era: digital city. Its main characteristic is the digitization of geographic information. The application is limited to some specialized agencies with limited clients. With the development of internet, broadband and wireless technologies, smart

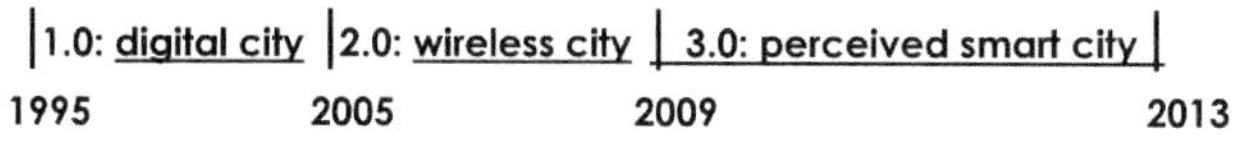

|1.0: <u>digital city</u> |2.0: <u>wireless city</u> | 3.0: <u>perceived smart city</u> |

1995         2005        2009                2013

**Figure 3.10**   Development of Chinese smart cities [22].

city construction has entered the 2.0 era in 2005: interconnected city or wireless city. Its main feature is the availability of full range informatization and interconnection. The scope of its application has extended to almost all industries, with clients covering governments, enterprises, residents and many more. However, the application is still relatively poor, and data isolation phenomenon is common.

Since the idea of smart earth was formally proposed in 2009–2011, the construction of China's smart city has entered the 3.0 era: perceived smart city. Its main feature is that IoT began to widely apply to a large number of front-end sensing and data acquiring. Also, 3G and Wi-Fi technologies are used for data transmitting. Cloud computing and big data technologies are used for back-end storage, processing and mining. The application scope and number of clients are more extensive than the 2.0 era. Since the twelfth Five-Year Plan, smart city has been regarded as a pioneer idea and exploration in promoting strategic new industries and urban informatization in China [23].

### 3.6.2 Three Models of Smart City Construction in China

Smart city is an integrated multi-level application system that is based on comprehensive information infrastructure and information platform, driven by government investment and resident needs [22]. As an effective means to integrate advanced technologies into urban meticulous management, smart city itself is a comprehensive application system covering all fields of new generation information technology. The construction of smart city is divided

Internet of Things  Information Infrastructure   Social Service

**Figure 3.11**   Typical three models of smart cities in China.

into early infrastructure construction, medium-term data processing facility construction, and late service platform construction. Most relevant external organizations involved are communication equipment manufacturers, system integrating enterprises, data acquiring and data analysing enterprises, communication operators and data service enterprises. The entire industry chain will benefit from smart city construction. There are three models of China's smart city construction.

- IoT industry development as driver

This model focuses on developing IoT related industries, providing policy support to IoT industries, constructing large-scale IoT industrial parks, attracting and cultivating researchers and scientists, and supporting a number of pilot enterprises and pilot projects. In short, it is a construction model that utilizes industrial development to promote applications in the society. Examples include Tianjin, Guangzhou, Hangzhou, Wuxi, Chengdu and more.

- Information infrastructure construction as pilot

This model focuses on building urban information infrastructure vigorously, establishing optical fibre networks, expanding home network cables, increasing wireless network coverage in public areas and network bandwidth, promoting the integration of the three networks, and installing wireless information equipment at large scales. The goal is to build an urban information network that is interconnected at any time and from anywhere. Examples include Shanghai, Nanjing, Fuzhou, Dalian and others.

- Social service and management application as breakthrough

This model focuses on constructing a number of social application demonstration projects in public safety, urban transportation, urban ecology and environment, logistics and supply chain, urban management and other fields, in order to build several demonstration bases. These key breakthroughs and demonstrations will gradually realize the in-depth construction of smart cities. Examples include Beijing, Ningbo, Wuhan, Shenyang, Shijiazhuang and other cities as well.

From the point of view of investment strategy, there are also three models of domestic smart city construction: government-leading model, operator-pushing model, and IT vendor-pushing model.

- Government-leading model

This model is characterized by establishing a working group that is led by government and involves leaders from all functional departments, so as to coordinate the information system construction comprehensively. Examples include Ningbo and Lianyungang.

- Operator-pushing model

This model is the most common one. Operators have provided an urban infrastructure for wire and wireless communications, and have extensive experience of data analysing and processing. They have also built long-term relationships with major IT vendors.

- IT vendor-pushing model

Current participants include IBM, Digital China, Yunsheng Technology, Insigma Technology and more. Particularly, IBM is the creator of the idea of smart city. It had connections to or strategic cooperation with more than 100 governmental organizations in the early period.

## 3.7 Trends of Smart City Development in China

- Internet+ becomes the new step forward in smart cities.

In the future, China will promote mobile internet, cloud computing and big data analysis, and IoT applications in modern manufacturing industry, in order to create a healthy environment for the development of e-commerce, industrial internet and internet finance, and to guide IT companies to explore the international market.

**Figure 3.12** Silk road economic belt.

- Regional cooperation is highlighted.

In order to promote the implementation of the "Belt and Road policy" and to rejuvenate the ancient Silk Road with vitality, the government of China has created and published *the Vision and Action for Promoting the Silk Road Economic Belt and the Maritime Silk Road in the 21st Century* [24]. China will establish collaborations with Asian-European countries in new forms and strengthen the communication and consultation with countries along the route, so that all parties can benefit from this mutual cooperation of new historical height. The stakeholders have promoted a list of key cooperating projects in infrastructure interconnections, industrial investment, resource development, financial cooperation, cultural exchange, ecological protection, and maritime cooperation.

- The social value of data mining continues to increase.

As China's information development deepens, a huge amount of industrial data has been accumulated and scattered in government departments, industrial platforms, enterprises and many other different entities. In the current era of Internet, various types of data will show a huge growth leading to the presence of the big data industry (so-called blowout trend). The popularization of various types of big data analysis and prediction application will effectively revitalize the accumulated data. Developing the potential value of data and eliminating data blind spots will provide decision makers with analysing, forecasting and other types of services, thereby making contributions to renewals in smart city construction. Soon, big data analysis will be widely applied in all fields. Government agencies will focus on overcoming internal barriers to enable smooth data sharing. This will also benefit big data mining and promote smart city construction.

- Mobile devices will bring great convenience to urban services.

Education, health care, social security, transportation and so on form the key criteria of a liveable city. These public services are closely related to residents' life and therefore will bring new challenges to governments, enterprises and social organizations in the future. Good collaboration will lead to good public services based on mobile internet technologies, and make life more convenient. Ordering food, booking taxis, e-learning, electronic medical care, and many other urban functions can be realized via mobile devices. Mobile internet has created favourable conditions for China's information economy to leap forward. The society can enjoy all kinds of public services in more and more convenient forms. Mobile internet has fully connected people and public services and substantially enhanced the overall efficiency and quality of social services. The promotion of mobile services will promote the informationization of urban life, and accelerate the development of transportation, medical care, environmental protection and public security. It is playing an important role in optimizing social resource allocation, innovating public service, improving service quality, and finally achieving information sharing in the whole society.

- Smart communities will become a most important new "playground" for smart city construction.

In recent years, all kinds of new technologies are rapidly growing and increasingly being applied to smart city construction. Smart community projects such as digital communities, smart home, community pension, and smart ecologic community are emerging one after another. With the promotion of smart city and the popularization of information technology, smart community projects are expecting a new round of rapid development. The competition of smart city construction heats up since 2015 due to model innovation, technology promotion and data accumulation. Smart communities are the physical bearer of many businesses, and therefore enterprises will accelerate their settlement in smart communities and push smart communities to achieve rapid expansion.

- Information security will become the new foundation of smart city construction.

During the construction of smart cities, infrastructure and information resources are important components of smart cities. Whether they can function well will directly affect the effectiveness of smart city. Information security is an auxiliary supporting system but of great importance in smart city construction. How to build an integrated information security monitoring platform and how to strengthen the information security risk assessment system will become strategic focuses of smart city construction. From the government side, it will promote the classification of infrastructure so as to continuously strengthen the security protection of network infrastructure and information resources.

- Smart city will emphasize the development of the information economy.

Information economy, as the new-born of the technology revolution, is going to accelerate its speed of growth with development of smart city construction. Based on virtual space, information economy will enjoy large potential in development. Safe, convenient and low-carbon e-commerce will be the main form in future economic activities of government, enterprises and individuals. The non-material network culture has broken the physical constraints of cultural carrier, content and communication. It has accelerated cultural diffusion and become an important component of information economy for many countries. The development of information economy will not contradict the existing agricultural economy, industrial economy, and service economy. On the contrary, it will improve the quality of these three economies through information technology, and lead to a situation where intangible information-based economy will replace the tangible material economy and become dominant. It is expected that in the coming years, information economy will further promote and drive the traditional economy to grow. Information economy will root deeply in the construction of smart city. Government will also be more willing to encourage and support traditional economy to restructure, and new information industry to develop with intelligence.

## Appendix 1: Abbreviations

[A]
APP   Application
[B]
BI   Business Intelligence
[C]
CVDS   Cardiovascular Diseases
[E]
ECG   ElectroCardioGram
EHR   Electronic health records
ETA   Estimated Time of Arrival
[G]
GIS   Geographic Information System
GPS   Global Positioning System
[I]
IaaS   Infrastructure-as-a-Service
ICF   Intelligent Community Forum
IPTV   Internet Protocol Television
ITS   Intelligent Transportation System
[M]
MIP   Mobile Instant Pages

| MRI | Magnetic resonance imaging |
| --- | --- |

[O]

| O2O | Online To Offline |
| --- | --- |
| OD | Origin Destination |

[P]

| PaaS | Platform-as-a-Service |
| --- | --- |
| POS | Point of Sale |

[R]

| RFID | Radio Frequency Identification |
| --- | --- |

[S]

| SAAS | Software-as-a-Service |
| --- | --- |
| SASAC | State Owned Assets Supervision and Administration |

[T]

| TOCC | Traffic Operation Control Center |
| --- | --- |

[U]

| UPS | United Parcel Service |
| --- | --- |

[W]

| WHO | World Health Organization |
| --- | --- |
| Wi-Fi | Wireless Fidelity |
| WiMAX | Worldwide Interoperability for Microwave Access |

# Appendix 2: Assessment of Smart Cities in China [25]

At present, smart city construction has become an irreversible trend of urban sustainable development in China and has brought solutions to urban development problems. As the new urbanization proceeds, Internet+, Online To Offline (O2O), internet finance, industry 4.0 and many other new opportunities, new models and new ideas have emerged. China's smart city construction is about to enter the golden period [26, 27]. Wuxi, Shanghai, Beijing, Hangzhou, Ningbo, Shenzhen, for example, have been leading the country for several years in terms of smart city construction.

We have created six categories (smart infrastructure, smart management, smart service, smart economy, smart population, and security system) plus an extra point to assess smart city development score. There are seven primary indicators and sixteen secondary indicators, and an extra six sub-items. Smart infrastructure includes infrastructure network construction, basic information resource construction and sharing, and urban cloud platform. Smart management includes e-government, public resource transaction platform, and social media participation. Smart service includes social service

**Table 3.2**  2015 smart cities in China development level assessment score [25]

| Rank | Secondary Indicator | Smart Infrastructure | | | | Smart Management | | Smart Service | | | Smart Economy | | Smart Population | | Security System | | | Extra | Score |
|---|---|---|---|---|---|---|---|---|---|---|---|---|---|---|---|---|---|---|---|
| | | Infrastructure Network Construction | Basic Information Resource Construction and Sharing | Urban Cloud Platform | e-government | Public Resource Transaction Platform | Social Media Participation | Social Service Level of Livelihood | Open Data Service Level | Urban Innovation and Start-Up Level | Economic Output Energy Consumption Level | Internet Industry Development Level | IT Service Industry Workers | Public Life Network Level | Planning and Development | Public Information Training | Performance Assessment | | |
| 1 | Wuxi | 4.70 | 9.50 | 4.50 | 9.00 | 3.00 | 3.73 | 6.00 | 7.50 | 3.50 | 2.00 | 3.44 | 2.00 | 3.74 | 4.80 | 4.00 | 4.00 | 4.80 | 80.20 |
| 2 | Shanghai | 4.58 | 8.50 | 4.00 | 6.00 | 1.50 | 4.80 | 7.50 | 7.50 | 4.60 | 2.00 | 3.69 | 3.50 | 3.28 | 4.80 | 5.00 | 4.00 | 4.80 | 80.05 |
| 3 | Beijing | 2.95 | 9.50 | 4.00 | 7.50 | 3.00 | 4.08 | 7.50 | 7.00 | 4.80 | 3.00 | 3.76 | 3.50 | 4.33 | 4.50 | 4.00 | 1.50 | 4.80 | 79.72 |
| 4 | Hangzhou | 3.36 | 9.50 | 4.50 | 8.00 | 4.50 | 4.33 | 7.00 | 3.50 | 4.20 | 2.50 | 4.14 | 2.50 | 4.65 | 4.50 | 1.00 | 4.00 | 4.80 | 76.97 |
| 5 | Ningbo | 4.80 | 9.50 | 4.50 | 8.50 | 4.50 | 3.83 | 6.00 | 4.50 | 2.70 | 2.00 | 3.17 | 2.00 | 3.87 | 4.80 | 3.00 | 4.50 | 4.50 | 76.67 |
| 6 | Shenzhen | 4.11 | 9.50 | 3.00 | 4.50 | 1.50 | 3.70 | 7.50 | 6.50 | 4.30 | 2.50 | 4.74 | 3.50 | 4.97 | 4.50 | 3.00 | 3.50 | 3.00 | 74.32 |
| 7 | Zhuhai | 2.39 | 7.00 | 2.50 | 9.00 | 4.50 | 1.85 | 7.00 | 3.00 | 2.60 | 2.50 | 2.80 | 3.00 | 4.34 | 3.00 | 3.00 | 3.00 | 2.50 | 63.97 |
| 8 | Foshan | 2.37 | 9.50 | 3.50 | 4.50 | 4.00 | 4.20 | 7.50 | 6.50 | 2.00 | 2.00 | 2.74 | 0.50 | 3.84 | 2.80 | 3.00 | 2.00 | 3.00 | 63.95 |
| 9 | Xiamen | 2.99 | 8.50 | 3.50 | 6.50 | 3.50 | 2.10 | 5.50 | 3.50 | 3.50 | 2.50 | 2.82 | 3.50 | 4.40 | 2.20 | 1.00 | 1.00 | 4.80 | 61.81 |
| 10 | Guangzhou | 4.14 | 9.50 | 3.50 | 4.50 | 4.50 | 2.40 | 7.50 | 3.00 | 4.10 | 2.00 | 3.51 | 2.50 | 2.31 | 2.50 | 2.50 | 0.50 | 2.50 | 61.45 |
| 11 | Qingdao | 4.43 | 8.50 | 1.10 | 4.00 | 4.00 | 3.93 | 5.50 | 3.00 | 2.60 | 3.50 | 1.77 | 2.00 | 3.53 | 4.00 | 3.00 | 1.00 | 4.80 | 60.65 |
| 12 | Nanjing | 2.95 | 9.50 | 1.50 | 3.50 | 3.50 | 4.25 | 4.00 | 2.00 | 3.80 | 2.00 | 3.06 | 3.50 | 4.06 | 4.00 | 3.00 | 2.00 | 4.00 | 60.62 |
| 13 | Suzhou | 2.84 | 9.50 | 4.00 | 4.50 | 3.00 | 3.53 | 4.00 | 2.00 | 3.70 | 1.50 | 3.74 | 2.00 | 4.10 | 3.00 | 1.00 | 3.00 | 3.50 | 58.90 |
| 14 | Jinhua | 2.85 | 7.70 | 3.10 | 8.50 | 4.50 | 2.10 | 6.00 | 3.50 | 1.00 | 1.50 | 3.80 | 1.50 | 3.85 | 2.30 | 3.00 | 1.00 | 2.00 | 58.20 |
| 15 | Chengdu | 1.62 | 9.50 | 1.50 | 5.50 | 4.00 | 3.80 | 7.00 | 2.00 | 4.10 | 2.00 | 2.34 | 2.50 | 3.74 | 2.60 | 1.00 | 1.50 | 2.50 | 57.20 |
| 16 | Wuhan | 3.09 | 6.50 | 3.50 | 3.50 | 4.00 | 3.73 | 7.00 | 7.50 | 3.50 | 1.00 | 2.50 | 1.00 | 3.01 | 2.00 | 0.50 | 0.50 | 4.00 | 56.83 |
| 17 | Hefei | 2.33 | 7.70 | 3.50 | 8.50 | 4.50 | 2.73 | 3.50 | 2.00 | 3.10 | 2.50 | 1.86 | 2.00 | 3.79 | 2.20 | 2.50 | 1.00 | 2.50 | 56.21 |
| 18 | Shaoxing | 2.94 | 9.50 | 1.70 | 7.50 | 4.00 | 2.50 | 6.00 | 3.50 | 0.60 | 2.00 | 1.91 | 1.00 | 3.53 | 3.00 | 2.00 | 2.50 | 1.50 | 55.68 |
| 19 | Jiaxing | 3.00 | 8.00 | 3.50 | 7.50 | 4.50 | 2.05 | 6.00 | 2.00 | 1.10 | 2.00 | 3.24 | 1.00 | 2.17 | 2.40 | 2.50 | 1.50 | 2.50 | 54.95 |
| 20 | Zhongshan | 2.20 | 9.50 | 1.50 | 9.00 | 4.00 | 1.30 | 7.00 | 2.00 | 1.30 | 1.00 | 2.84 | 1.50 | 4.12 | 2.20 | 3.00 | 1.00 | 1.00 | 54.46 |
| 21 | Wenzhou | 3.55 | 5.00 | 3.00 | 8.00 | 4.00 | 3.15 | 6.00 | 2.00 | 1.20 | 2.50 | 3.13 | 1.00 | 3.65 | 2.50 | 1.50 | 1.00 | 3.00 | 54.17 |
| 22 | Weifang | 4.11 | 9.50 | 2.00 | 7.50 | 4.00 | 3.65 | 4.00 | 3.00 | 0.70 | 1.00 | 0.98 | 1.50 | 3.11 | 1.50 | 0.50 | 3.50 | 3.00 | 53.54 |
| 23 | Lishui | 1.75 | 8.50 | 1.50 | 8.00 | 4.00 | 2.85 | 6.00 | 3.50 | 0.70 | 2.50 | 2.71 | 1.50 | 2.39 | 2.80 | 1.00 | 0.50 | 3.00 | 53.19 |
| 24 | Dongguan | 2.39 | 7.70 | 3.50 | 4.50 | 4.00 | 2.20 | 7.00 | 0.50 | 2.00 | 2.00 | 2.77 | 1.50 | 4.07 | 2.50 | 2.00 | 1.50 | 2.50 | 52.63 |
| 25 | Taizhou | 2.56 | 8.00 | 1.00 | 7.50 | 4.00 | 3.00 | 6.00 | 3.50 | 1.00 | 1.00 | 1.98 | 1.00 | 3.29 | 2.50 | 1.00 | 1.50 | 3.00 | 51.83 |
| 26 | Tianjin | 1.68 | 6.50 | 3.50 | 6.50 | 3.00 | 2.10 | 1.00 | 3.50 | 3.10 | 2.00 | 2.19 | 1.50 | 3.23 | 4.00 | 3.00 | 1.50 | 3.00 | 51.29 |
| 27 | Huizhou | 2.30 | 6.70 | 3.10 | 8.50 | 4.00 | 2.90 | 7.00 | 1.00 | 0.70 | 1.00 | 1.79 | 1.50 | 3.59 | 2.50 | 1.00 | 2.50 | 1.00 | 51.08 |
| 28 | Zhengzhou | 2.12 | 7.20 | 1.60 | 7.50 | 1.50 | 2.23 | 4.50 | 0.50 | 3.10 | 1.00 | 2.92 | 1.00 | 3.84 | 2.10 | 3.00 | 2.50 | 4.00 | 50.60 |

| | | | | | | | | | | | | | | | | | | |
|---|---|---|---|---|---|---|---|---|---|---|---|---|---|---|---|---|---|---|
| 29 | Huzhou | 2.91 | 9.50 | 1.10 | 8.00 | 4.00 | 2.78 | 6.00 | 3.50 | 0.60 | 1.00 | 1.92 | 0.50 | 3.51 | 1.60 | 0.50 | 0.50 | 2.00 | 49.91 |
| 30 | Shantou | 1.58 | 4.50 | 2.00 | 9.00 | 2.50 | 1.10 | 7.00 | 6.50 | 0.80 | 2.00 | 1.70 | 0.50 | 3.34 | 2.20 | 2.50 | 1.00 | 1.00 | 49.22 |
| 31 | Zhoushan | 2.45 | 7.20 | 2.50 | 8.00 | 1.50 | 2.55 | 6.00 | 2.00 | 0.50 | 4.00 | 1.71 | 1.00 | 3.62 | 3.00 | 1.00 | 2.00 | 0.50 | 49.02 |
| 32 | Yichang | 1.42 | 9.50 | 2.50 | 4.50 | 4.50 | 3.08 | 4.00 | 2.00 | 0.90 | 1.00 | 0.90 | 1.50 | 2.94 | 2.20 | 1.00 | 4.50 | 2.50 | 48.94 |
| 33 | Nanchang | 2.13 | 9.50 | 2.50 | 1.00 | 4.50 | 3.78 | 4.50 | 0.50 | 2.00 | 2.00 | 2.79 | 2.00 | 3.69 | 2.50 | 1.00 | 0.50 | 4.00 | 48.89 |
| 34 | Quanzhou | 3.05 | 7.00 | 2.00 | 8.50 | 4.00 | 0.00 | 4.50 | 1.00 | 0.90 | 2.00 | 2.04 | 1.50 | 3.84 | 2.00 | 3.00 | 1.50 | 2.00 | 48.83 |
| 35 | Changsha | 2.02 | 4.40 | 2.50 | 7.00 | 4.00 | 0.50 | 6.00 | 2.00 | 3.20 | 2.00 | 1.88 | 2.00 | 3.82 | 1.70 | 1.00 | 1.00 | 3.00 | 48.02 |
| 36 | Fuzhou | 3.45 | 4.00 | 1.50 | 3.50 | 4.00 | 1.30 | 5.50 | 3.50 | 3.10 | 2.50 | 2.39 | 1.00 | 2.27 | 1.80 | 1.00 | 2.00 | 4.50 | 47.31 |
| 37 | Jiangmen | 2.18 | 7.70 | 1.50 | 6.50 | 4.00 | 2.80 | 7.00 | 1.00 | 0.50 | 2.00 | 1.16 | 0.50 | 3.31 | 2.50 | 2.50 | 0.50 | 1.50 | 47.15 |
| 38 | Changzhou | 2.70 | 9.50 | 1.50 | 5.50 | 2.00 | 1.60 | 1.00 | 2.00 | 2.60 | 1.50 | 0.92 | 2.00 | 3.67 | 2.50 | 2.50 | 1.50 | 2.00 | 46.99 |
| 39 | Xi'an | 3.03 | 9.50 | 1.80 | 1.00 | 2.50 | 3.38 | 0.50 | 2.00 | 3.50 | 2.50 | 1.81 | 3.50 | 3.82 | 1.50 | 2.00 | 0.50 | 4.00 | 46.83 |
| 40 | Xiangyang | 1.35 | 9.50 | 3.00 | 8.50 | 4.50 | 0.20 | 4.00 | 1.00 | 0.50 | 1.00 | 2.33 | 0.50 | 2.44 | 3.00 | 0.50 | 1.50 | 3.00 | 46.82 |
| 41 | Yangzhou | 2.10 | 9.50 | 1.50 | 5.50 | 4.00 | 1.50 | 0.50 | 2.00 | 1.30 | 2.00 | 1.69 | 2.00 | 3.36 | 2.50 | 2.50 | 1.50 | 3.00 | 46.46 |
| 42 | Guiyang | 1.96 | 5.00 | 2.50 | 9.00 | 1.50 | 2.18 | 3.50 | 2.00 | 2.00 | 1.00 | 1.13 | 2.00 | 3.54 | 2.10 | 0.50 | 2.50 | 4.00 | 46.41 |
| 43 | Nanping | 2.31 | 7.50 | 1.70 | 8.50 | 4.00 | 0.90 | 4.00 | 3.50 | 0.60 | 1.00 | 1.04 | 0.50 | 3.05 | 2.00 | 1.00 | 1.50 | 3.00 | 46.10 |
| 44 | Harbin | 1.41 | 9.50 | 3.00 | 5.50 | 2.50 | 3.48 | 1.00 | 2.00 | 2.40 | 1.00 | 1.45 | 2.50 | 3.16 | 2.00 | 2.00 | 1.00 | 2.00 | 45.90 |
| 45 | Yinchuan | 1.65 | 9.50 | 1.60 | 3.50 | 4.00 | 2.85 | 4.00 | 1.00 | 0.50 | 1.00 | 2.08 | 1.50 | 3.49 | 1.60 | 0.50 | 2.00 | 4.80 | 45.57 |
| 46 | Quzhou | 1.04 | 3.60 | 1.20 | 8.00 | 4.50 | 2.80 | 6.00 | 2.00 | 0.50 | 1.00 | 1.09 | 2.50 | 3.21 | 2.00 | 2.00 | 1.50 | 2.50 | 45.43 |
| 47 | Weihai | 2.40 | 6.70 | 2.50 | 8.00 | 2.50 | 2.70 | 0.50 | 2.00 | 0.60 | 2.00 | 1.16 | 1.00 | 2.50 | 1.80 | 0.50 | 4.00 | 4.00 | 44.85 |
| 48 | Chongqing | 1.99 | 6.50 | 3.00 | 1.00 | 2.50 | 3.65 | 1.00 | 4.00 | 3.40 | 3.50 | 0.95 | 1.50 | 2.34 | 4.50 | 3.00 | 1.00 | 1.00 | 44.84 |
| 49 | Dalian | 2.12 | 4.20 | 2.50 | 5.00 | 4.00 | 0.00 | 0.50 | 2.00 | 2.70 | 1.00 | 2.16 | 2.50 | 3.49 | 3.70 | 2.50 | 1.50 | 4.80 | 44.66 |
| 50 | Putian | 1.94 | 6.50 | 1.50 | 8.50 | 4.50 | 0.00 | 4.00 | 0.50 | 0.50 | 2.50 | 2.72 | 0.50 | 4.02 | 1.70 | 1.00 | 1.00 | 3.00 | 44.39 |
| 51 | Xinyu | 1.49 | 9.50 | 1.10 | 5.50 | 4.00 | 3.93 | 4.50 | 0.50 | 0.60 | 1.00 | 1.42 | 0.50 | 2.88 | 2.20 | 2.00 | 2.00 | 1.00 | 44.11 |
| 52 | Taiyuan | 2.64 | 4.70 | 2.80 | 5.00 | 4.00 | 0.40 | 4.50 | 2.00 | 2.00 | 1.00 | 2.27 | 3.00 | 3.78 | 2.20 | 1.00 | 2.00 | 0.50 | 43.79 |
| 53 | Wuhu | 1.47 | 3.60 | 2.50 | 4.00 | 2.00 | 2.38 | 4.50 | 3.50 | 0.90 | 2.00 | 1.67 | 3.50 | 3.09 | 2.00 | 2.50 | 1.00 | 3.00 | 43.61 |
| 54 | Jinan | 4.65 | 2.40 | 1.50 | 3.00 | 4.00 | 1.20 | 4.00 | 1.00 | 2.50 | 1.50 | 2.32 | 3.50 | 3.65 | 1.60 | 2.50 | 1.00 | 2.50 | 42.83 |
| 55 | Jiujiang | 1.26 | 7.70 | 1.50 | 7.00 | 4.00 | 2.13 | 0.50 | 0.50 | 0.70 | 1.50 | 1.86 | 1.50 | 2.19 | 2.30 | 2.50 | 2.00 | 3.00 | 42.14 |
| 56 | Zhanjiang | 1.13 | 7.50 | 1.30 | 9.50 | 1.50 | 0.20 | 7.00 | 5.00 | 0.50 | 1.00 | 1.80 | 0.50 | 2.59 | 1.50 | 0.50 | 0.50 | 0.00 | 42.02 |
| 57 | Longyan | 1.80 | 5.00 | 1.50 | 9.00 | 4.00 | 0.00 | 4.00 | 3.00 | 0.70 | 1.50 | 1.01 | 0.50 | 3.26 | 2.20 | 0.50 | 1.00 | 3.00 | 41.96 |
| 58 | Nanyang | 0.65 | 7.00 | 1.30 | 7.50 | 4.50 | 1.60 | 4.00 | 0.50 | 0.50 | 2.50 | 1.26 | 1.50 | 2.25 | 2.00 | 2.00 | 0.50 | 2.00 | 41.57 |
| 59 | Liupanshui | 0.63 | 9.50 | 0.50 | 9.50 | 4.00 | 1.50 | 3.50 | 0.50 | 0.50 | 1.00 | 1.71 | 0.50 | 2.61 | 1.00 | 2.50 | 0.50 | 1.00 | 40.94 |
| 60 | Lanzhou | 1.67 | 9.50 | 2.20 | 5.00 | 4.50 | 0.90 | 0.50 | 0.50 | 1.00 | 1.00 | 1.54 | 2.50 | 3.42 | 2.30 | 0.50 | 1.00 | 2.50 | 40.53 |

*(Continued)*

**Table 3.2**  Continued

| Primary Indicator | | Smart Infrastructure | | | | Smart Management | | Smart Service | | | Smart Economy | | Smart Population | | | Security System | | | |
| --- | --- | --- | --- | --- | --- | --- | --- | --- | --- | --- | --- | --- | --- | --- | --- | --- | --- | --- | --- |
| Rank | Secondary Indicator | Infrastructure Network Construction | Basic Information Resource Construction and Sharing | Urban Cloud Platform | e-government | Public Resource Transaction Platform | Social Media Participation | Social Service Level of Livelihood | Open Data Service Level | Urban Innovation and Start-Up Level | Economic Output Energy Consumption Level | Internet Industry Development Level | IT Service Industry Workers | Public Life Network Level | Planning and Development | Public Information Training | Performance Assessment | Extra | Score |
| 61 | Huanggang | 1.06 | 6.70 | 1.80 | 6.50 | 4.00 | 0.00 | 3.50 | 3.50 | 0.50 | 1.50 | 1.26 | 1.50 | 0.55 | 2.50 | 1.00 | 2.50 | 2.00 | 40.37 |
| 62 | Nanning | 1.91 | 6.70 | 2.10 | 2.00 | 1.50 | 3.58 | 0.50 | 0.50 | 2.00 | 2.50 | 1.60 | 3.50 | 1.52 | 2.00 | 2.50 | 3.00 | 2.50 | 39.90 |
| 63 | Shenzhou | 0.88 | 7.00 | 1.50 | 9.50 | 1.50 | 2.13 | 0.50 | 0.50 | 0.60 | 1.00 | 2.18 | 2.00 | 2.80 | 1.50 | 3.00 | 1.00 | 2.00 | 39.58 |
| 64 | Luoyang | 1.58 | 3.60 | 2.10 | 5.00 | 1.50 | 3.85 | 3.50 | 2.00 | 1.10 | 1.50 | 1.43 | 2.00 | 2.20 | 2.00 | 2.00 | 2.50 | 1.50 | 39.36 |
| 65 | Changchun | 1.90 | 8.50 | 2.00 | 3.00 | 1.00 | 1.60 | 0.50 | 1.00 | 0.90 | 1.50 | 2.00 | 2.50 | 3.26 | 3.60 | 2.50 | 1.50 | 2.00 | 39.26 |
| 66 | Hohhot | 0.35 | 4.50 | 2.00 | 3.00 | 4.50 | 1.90 | 0.50 | 2.00 | 1.50 | 1.00 | 2.59 | 1.00 | 1.27 | 4.00 | 2.50 | 1.50 | 4.80 | 38.91 |
| 67 | Shenyang | 1.74 | 4.20 | 2.00 | 3.00 | 1.50 | 2.20 | 0.50 | 3.00 | 2.40 | 1.50 | 2.13 | 1.00 | 3.40 | 3.80 | 2.50 | 4.00 | 4.00 | 38.86 |
| 68 | Kunming | 2.23 | 5.00 | 0.90 | 1.00 | 4.00 | 3.08 | 3.50 | 0.50 | 2.50 | 1.00 | 1.15 | 3.50 | 3.48 | 2.00 | 1.50 | 1.00 | 3.00 | 38.83 |
| 69 | Taizhou | 1.24 | 7.20 | 3.00 | 4.50 | 3.00 | 1.60 | 0.50 | 2.00 | 0.60 | 2.00 | 1.63 | 2.50 | 2.95 | 2.00 | 1.00 | 0.50 | 2.50 | 38.71 |
| 70 | Zibo | 2.14 | 5.20 | 1.00 | 3.00 | 4.50 | 0.70 | 4.00 | 2.00 | 0.50 | 1.00 | 1.06 | 0.50 | 3.18 | 2.80 | 3.00 | 2.50 | 1.50 | 38.58 |
| 71 | Bozhou | 0.55 | 7.20 | 2.10 | 7.50 | 4.00 | 1.93 | 0.50 | 0.50 | 0.50 | 2.00 | 1.27 | 1.50 | 1.66 | 2.20 | 1.00 | 1.50 | 2.50 | 38.40 |
| 72 | Fuyang | 0.57 | 9.50 | 1.10 | 4.50 | 4.00 | 3.00 | 0.50 | 2.00 | 0.60 | 1.00 | 1.26 | 1.00 | 1.87 | 2.30 | 0.50 | 2.00 | 2.00 | 37.70 |
| 73 | Xiangtan | 1.05 | 5.20 | 0.50 | 7.50 | 2.50 | 0.00 | 0.50 | 2.00 | 0.80 | 1.00 | 2.43 | 0.50 | 2.90 | 2.80 | 2.50 | 3.00 | 2.50 | 37.68 |
| 74 | Changde | 1.01 | 7.00 | 1.20 | 5.50 | 4.50 | 2.20 | 0.50 | 2.00 | 0.60 | 2.00 | 0.77 | 0.50 | 1.91 | 2.30 | 1.00 | 3.00 | 1.00 | 36.99 |
| 75 | Hengyang | 0.72 | 9.50 | 1.10 | 9.00 | 1.50 | 2.55 | 0.50 | 2.00 | 0.60 | 1.00 | 1.30 | 0.50 | 1.82 | 1.80 | 1.00 | 1.00 | 1.00 | 36.90 |
| 76 | Mianyang | 1.33 | 7.00 | 2.50 | 3.00 | 4.00 | 2.15 | 0.50 | 2.00 | 0.70 | 1.00 | 1.88 | 1.00 | 2.71 | 1.60 | 2.50 | 1.50 | 1.50 | 36.87 |
| 77 | Zhenjiang | 2.13 | 4.70 | 3.30 | 2.50 | 4.00 | 2.90 | 0.50 | 2.00 | 1.00 | 2.00 | 1.73 | 3.00 | 1.71 | 2.20 | 1.00 | 0.50 | 1.50 | 36.67 |
| 78 | Tianshui | 0.10 | 7.00 | 1.40 | 9.00 | 4.50 | 1.58 | 0.50 | 2.00 | 0.50 | 1.00 | 0.67 | 0.50 | 2.20 | 2.50 | 2.50 | 0.50 | 0.00 | 36.45 |
| 79 | Tangshan | 1.36 | 5.10 | 1.70 | 5.50 | 2.50 | 1.10 | 0.50 | 3.00 | 0.60 | 1.00 | 2.45 | 3.00 | 1.62 | 1.50 | 1.00 | 0.50 | 4.00 | 36.43 |
| 80 | Yantai | 2.35 | 4.40 | 1.50 | 2.00 | 2.50 | 2.98 | 4.00 | 2.00 | 1.00 | 2.00 | 1.55 | 1.50 | 1.82 | 2.80 | 1.00 | 1.00 | 2.00 | 36.40 |
| 81 | Haikou | 2.35 | 4.70 | 1.30 | 2.50 | 1.00 | 0.00 | 4.00 | 2.00 | 1.50 | 2.00 | 2.26 | 2.50 | 3.75 | 2.00 | 0.50 | 1.00 | 2.50 | 35.85 |
| 82 | Zhuzhou | 1.27 | 7.00 | 3.50 | 5.50 | 1.00 | 2.35 | 0.50 | 3.50 | 0.60 | 1.50 | 1.51 | 0.50 | 0.79 | 2.30 | 0.50 | 1.00 | 2.50 | 35.82 |
| 83 | Zhangzhou | 0.67 | 2.70 | 2.00 | 9.00 | 4.00 | 0.00 | 4.00 | 0.50 | 0.50 | 1.00 | 0.97 | 0.50 | 3.15 | 1.80 | 1.00 | 1.00 | 3.00 | 35.79 |
| 84 | Linyi | 1.43 | 7.00 | 0.90 | 1.00 | 4.50 | 0.50 | 3.50 | 2.00 | 0.50 | 1.00 | 0.91 | 2.00 | 0.64 | 2.70 | 3.00 | 1.00 | 3.00 | 35.58 |
| 85 | Maanshan | 1.41 | 7.00 | 2.00 | 4.50 | 4.50 | 1.10 | 0.50 | 2.00 | 0.70 | 2.00 | 1.08 | 0.50 | 2.78 | 2.00 | 1.00 | 1.00 | 2.50 | 35.57 |
| 86 | Huaian | 1.07 | 9.50 | 2.80 | 1.50 | 4.00 | 1.10 | 0.50 | 2.00 | 0.70 | 1.00 | 0.95 | 1.50 | 2.14 | 1.80 | 1.00 | 1.50 | 2.50 | 35.56 |
| 87 | Jingzhou | 1.72 | 4.00 | 1.70 | 5.50 | 4.00 | 0.00 | 4.00 | 2.00 | 0.50 | 1.00 | 0.84 | 2.00 | 2.10 | 2.20 | 1.00 | 1.00 | 2.00 | 35.56 |
| 88 | Yuxi | 1.00 | 5.00 | 1.90 | 2.50 | 4.00 | 2.18 | 5.50 | 2.00 | 0.50 | 1.50 | 0.82 | 0.50 | 2.71 | 1.60 | 0.50 | 1.00 | 1.50 | 34.71 |
| 89 | Jining | 1.39 | 7.70 | 1.40 | 1.50 | 4.00 | 1.50 | 4.00 | 2.00 | 0.50 | 1.00 | 0.92 | 1.00 | 2.79 | 1.40 | 0.50 | 0.50 | 2.50 | 34.60 |

| | | | | | | | | | | | | | | | | | | | |
|---|---|---|---|---|---|---|---|---|---|---|---|---|---|---|---|---|---|---|---|
| 90 | Dezhou | 1.14 | 5.40 | 2.50 | 1.50 | 4.50 | 1.60 | 4.00 | 0.50 | 0.50 | 1.00 | 0.85 | 0.50 | 2.54 | 1.50 | 2.50 | 2.00 | 2.00 | 34.54 |
| 91 | Changzhi | 1.31 | 4.40 | 0.50 | 6.00 | 4.00 | 0.00 | 4.50 | 0.50 | 0.50 | 1.00 | 1.31 | 2.00 | 2.74 | 1.60 | 2.00 | 0.50 | 1.50 | 34.36 |
| 92 | Nantong | 2.07 | 4.70 | 2.50 | 1.50 | 4.50 | 1.30 | 0.50 | 2.00 | 1.40 | 2.50 | 1.69 | 1.50 | 1.00 | 2.00 | 0.50 | 0.50 | 2.50 | 34.16 |
| 93 | Liuzhou | 1.14 | 7.70 | 1.90 | 2.50 | 2.00 | 3.80 | 0.50 | 2.00 | 0.60 | 1.00 | 2.43 | 2.00 | 0.85 | 2.50 | 0.50 | 1.50 | 1.00 | 33.92 |
| 94 | Sanya | 0.10 | 4.50 | 1.70 | 9.50 | 1.50 | 2.68 | 4.00 | 0.50 | 0.70 | 1.00 | 1.44 | 2.00 | 1.69 | 1.00 | 0.50 | 0.50 | 0.50 | 33.80 |
| 95 | Yingtan | 1.25 | 9.50 | 0.50 | 2.50 | 4.00 | 0.20 | 1.00 | 2.00 | 0.50 | 2.50 | 0.91 | 0.50 | 2.33 | 2.30 | 0.50 | 1.50 | 1.50 | 33.50 |
| 96 | Yancheng | 1.40 | 4.50 | 3.00 | 2.00 | 4.50 | 0.00 | 0.50 | 2.00 | 0.60 | 2.00 | 1.54 | 1.00 | 2.92 | 2.50 | 1.00 | 1.00 | 1.50 | 33.45 |
| 97 | Ordos | 0.97 | 7.00 | 1.50 | 3.50 | 4.00 | 3.05 | 0.50 | 0.50 | 0.70 | 1.00 | 0.97 | 2.00 | 1.73 | 1.50 | 1.00 | 1.00 | 1.00 | 33.42 |
| 98 | Yongzhou | 1.31 | 7.00 | 1.10 | 6.50 | 4.50 | 2.63 | 0.50 | 2.00 | 0.50 | 1.00 | 0.77 | 0.50 | 1.92 | 1.60 | 1.00 | 0.50 | 1.00 | 33.34 |
| 99 | Baotou | 0.96 | 5.70 | 1.50 | 4.50 | 2.50 | 1.80 | 0.50 | 2.00 | 0.60 | 1.00 | 0.92 | 3.50 | 3.17 | 2.00 | 1.00 | 1.50 | 1.00 | 33.15 |
| 100 | Huainan | 1.34 | 6.20 | 1.00 | 6.50 | 4.00 | 2.10 | 0.50 | 0.50 | 0.60 | 1.00 | 0.90 | 0.50 | 2.40 | 2.00 | 1.00 | 1.50 | 1.00 | 33.04 |
| 101 | Yichun | 1.01 | 8.00 | 1.00 | 3.00 | 4.00 | 1.50 | 0.50 | 0.50 | 0.60 | 2.00 | 1.79 | 3.00 | 1.90 | 2.20 | 0.50 | 0.50 | 2.00 | 33.00 |
| 102 | Hulunbuir | 1.24 | 4.30 | 1.00 | 3.50 | 4.00 | 3.78 | 0.50 | 3.00 | 0.60 | 1.00 | 0.79 | 0.50 | 2.97 | 1.20 | 3.00 | 1.00 | 1.00 | 32.88 |
| 103 | Zhunyi | 0.28 | 5.50 | 2.70 | 1.50 | 4.50 | 0.00 | 3.50 | 0.50 | 0.50 | 1.00 | 1.25 | 1.50 | 2.59 | 2.80 | 0.50 | 1.50 | 0.50 | 32.63 |
| 104 | Lianyungang | 1.81 | 5.20 | 2.30 | 2.50 | 4.00 | 0.30 | 0.50 | 2.00 | 0.70 | 1.00 | 1.57 | 3.00 | 2.06 | 1.60 | 2.50 | 0.50 | 2.00 | 32.55 |
| 105 | Xuzhou | 0.33 | 2.40 | 2.30 | 2.50 | 2.00 | 3.05 | 1.00 | 2.00 | 2.50 | 1.00 | 1.08 | 1.00 | 3.17 | 2.60 | 1.00 | 1.50 | 3.00 | 32.43 |
| 106 | Baoji | 1.38 | 6.10 | 1.60 | 3.00 | 2.50 | 0.10 | 0.50 | 0.50 | 0.50 | 1.50 | 0.80 | 3.00 | 2.70 | 1.70 | 2.00 | 2.00 | 2.50 | 31.89 |
| 107 | Anshan | 1.67 | 1.40 | 1.50 | 2.00 | 4.50 | 4.05 | 1.00 | 0.50 | 0.60 | 1.00 | 0.89 | 0.50 | 3.03 | 1.50 | 2.00 | 2.00 | 3.00 | 31.65 |
| 108 | Shijiazhuang | 1.07 | 3.00 | 3.00 | 2.00 | 1.50 | 3.83 | 0.50 | 2.00 | 1.50 | 1.00 | 1.09 | 2.00 | 1.03 | 3.00 | 1.50 | 0.50 | 3.50 | 31.51 |
| 109 | Urunqi | 2.13 | 4.70 | 1.10 | 2.50 | 1.50 | 3.20 | 0.50 | 1.00 | 1.00 | 1.00 | 1.15 | 2.00 | 3.47 | 1.80 | 1.00 | 2.20 | 2.00 | 31.25 |
| 110 | Langfang | 0.63 | 5.00 | 2.50 | 3.00 | 1.50 | 0.00 | 0.50 | 3.00 | 1.30 | 1.50 | 2.13 | 1.50 | 3.30 | 1.50 | 0.50 | 1.50 | 3.00 | 30.86 |
| 111 | Ji'an | 0.15 | 6.70 | 1.80 | 3.00 | 4.00 | 1.10 | 0.50 | 2.00 | 0.50 | 1.00 | 0.80 | 0.50 | 1.80 | 2.50 | 0.50 | 0.50 | 2.50 | 30.85 |
| 112 | Zhaoqing | 0.10 | 6.70 | 1.50 | 4.00 | 4.00 | 0.00 | 7.00 | 1.00 | 0.60 | 1.00 | 0.93 | 0.50 | 0.75 | 1.60 | 0.50 | 0.50 | 0.50 | 30.68 |
| 113 | Guilin | 1.54 | 4.20 | 2.00 | 2.00 | 2.50 | 2.60 | 0.50 | 1.30 | 1.50 | 1.00 | 0.91 | 1.00 | 2.61 | 3.00 | 0.20 | 1.90 | 4.80 | 30.45 |
| 114 | Karamay | 2.24 | 9.50 | 2.00 | 1.00 | 1.50 | 1.10 | 0.50 | 1.00 | 0.50 | 1.00 | 1.11 | 0.50 | 1.29 | 1.60 | 2.00 | 0.50 | 0.50 | 30.34 |
| 115 | Bengbu | 1.11 | 2.70 | 2.30 | 1.50 | 4.50 | 1.10 | 0.50 | 1.00 | 0.60 | 2.00 | 1.41 | 1.50 | 2.39 | 1.70 | 2.50 | 0.50 | 1.50 | 30.31 |
| 116 | Dali | 0.10 | 4.00 | 2.00 | 4.00 | 4.50 | 2.60 | 1.00 | 1.00 | 0.50 | 1.00 | 1.01 | 0.50 | 0.99 | 2.00 | 1.00 | 3.00 | 3.00 | 30.20 |
| 117 | Fuzhou | 0.19 | 6.90 | 1.10 | 3.00 | 4.00 | 3.25 | 0.50 | 0.50 | 0.50 | 2.50 | 0.81 | 0.50 | 0.61 | 1.80 | 1.50 | 0.50 | 0.50 | 30.15 |
| 118 | Datong | 1.58 | 2.50 | 1.60 | 2.50 | 2.50 | 0.00 | 4.50 | 4.50 | 0.50 | 1.00 | 0.83 | 1.50 | 3.03 | 2.60 | 1.00 | 3.00 | 3.00 | 30.14 |
| 119 | Qinhuangdao | 1.57 | 6.00 | 1.50 | 2.50 | 4.50 | 2.35 | 0.50 | 0.50 | 0.60 | 1.00 | 1.00 | 0.50 | 1.80 | 2.20 | 1.50 | 0.50 | 0.50 | 30.02 |
| 120 | Baoding | 1.14 | 5.50 | 1.50 | 5.00 | 4.00 | 2.15 | 0.50 | 0.50 | 0.70 | 1.50 | 0.92 | 1.50 | 1.10 | 1.50 | 1.00 | 0.50 | 0.50 | 30.01 |
| 121 | Chuzhou | 1.14 | 4.40 | 1.00 | 2.00 | 4.00 | 2.48 | 0.50 | 2.00 | 0.50 | 1.00 | 0.85 | 0.50 | 2.32 | 1.80 | 2.50 | 1.00 | 2.00 | 29.98 |
| 122 | Shangrao | 0.60 | 9.50 | 2.50 | 4.00 | 2.00 | 0.40 | 1.00 | 0.50 | 0.50 | 2.50 | 0.81 | 0.50 | 0.56 | 2.10 | 0.50 | 0.50 | 1.00 | 29.97 |
| 123 | Tongren | 0.65 | 2.70 | 1.90 | 4.00 | 4.00 | 2.13 | 3.50 | 0.50 | 0.50 | 1.00 | 0.70 | 0.50 | 2.02 | 2.00 | 1.50 | 1.50 | 0.50 | 29.60 |
| 124 | Xianyang | 1.10 | 4.40 | 2.80 | 1.50 | 1.00 | 0.88 | 2.50 | 0.50 | 0.50 | 1.00 | 0.79 | 0.50 | 2.68 | 2.30 | 2.50 | 1.50 | 2.50 | 28.95 |

*(Continued)*

**Table 3.2**  Continued

| Primary Indicator | | Smart Infrastructure | | | | Smart Management | | Smart Service | | | Smart Economy | | Smart Population | | Security System | | Performance | | |
|---|---|---|---|---|---|---|---|---|---|---|---|---|---|---|---|---|---|---|---|
| Secondary Indicator | | Infrastructure Network Construction | Basic Information Resource Construction and Sharing | Urban Cloud Platform | e-government | Public Resource Transaction Platform | Social Media Participation | Social Service Level of Livelihood | Open Data Service Level | Urban Innovation and Start-Up Level | Economic Output Energy Consumption Level | Internet Industry Development Level | IT Service Industry Workers | Public Life Network Level | Planning and Development | Public Information Training | Assessment | Extra | Score |
| Rank | | | | | | | | | | | | | | | | | | | |
| 125 | Handan | 0.85 | 1.60 | 1.50 | 5.50 | 1.50 | 2.23 | 0.50 | 2.00 | 0.60 | 1.00 | 1.32 | 1.50 | 2.34 | 1.50 | 0.50 | 1.50 | 3.00 | 28.94 |
| 126 | Xining | 1.46 | 4.40 | 2.50 | 3.50 | 1.00 | 0.30 | 0.50 | 0.50 | 0.50 | 1.00 | 0.92 | 1.50 | 3.14 | 1.50 | 2.50 | 1.00 | 2.50 | 28.71 |
| 127 | Lasa | 0.10 | 2.80 | 1.50 | 6.00 | 4.00 | 1.93 | 1.00 | 0.50 | 0.50 | 1.00 | 1.05 | 1.50 | 3.35 | 2.10 | 0.50 | 0.50 | 0.00 | 28.32 |
| 128 | Yibin | 0.61 | 4.70 | 1.80 | 2.50 | 4.00 | 2.35 | 0.50 | 2.00 | 0.50 | 1.00 | 0.79 | 0.50 | 1.55 | 1.80 | 0.20 | 1.00 | 2.50 | 28.30 |
| 129 | Cangzhou | 0.10 | 4.40 | 1.00 | 5.50 | 4.00 | 2.65 | 1.00 | 0.50 | 0.50 | 1.00 | 0.92 | 0.50 | 0.71 | 1.50 | 0.50 | 2.00 | 1.00 | 27.78 |
| 130 | Daqing | 1.51 | 6.00 | 1.00 | 2.00 | 3.00 | 2.13 | 0.50 | 2.00 | 0.50 | 1.00 | 0.94 | 0.50 | 0.90 | 3.80 | 0.50 | 1.00 | 0.50 | 27.77 |
| 131 | Lijiang | 1.07 | 3.00 | 0.50 | 2.50 | 4.00 | 1.93 | 3.50 | 0.50 | 0.50 | 1.00 | 1.40 | 0.50 | 2.43 | 1.80 | 1.00 | 0.50 | 1.50 | 27.62 |
| 132 | Yanan | 1.86 | 3.20 | 1.20 | 4.50 | 1.00 | 0.50 | 0.50 | 3.50 | 0.50 | 1.00 | 0.81 | 0.50 | 3.02 | 2.30 | 1.00 | 1.00 | 1.00 | 27.38 |
| 133 | Huangshan | 1.23 | 3.00 | 1.50 | 3.50 | 4.50 | 1.25 | 0.50 | 0.50 | 0.50 | 1.00 | 0.96 | 0.50 | 2.61 | 1.80 | 0.50 | 0.50 | 3.00 | 27.35 |
| 134 | Yingkou | 1.50 | 7.00 | 1.50 | 1.50 | 4.50 | 0.00 | 0.50 | 0.50 | 0.50 | 1.00 | 0.89 | 0.50 | 3.07 | 1.70 | 0.50 | 0.50 | 1.50 | 27.16 |
| 135 | Kaifeng | 0.89 | 4.70 | 1.00 | 3.00 | 4.00 | 0.90 | 1.00 | 0.50 | 0.50 | 1.00 | 0.80 | 0.50 | 2.26 | 1.50 | 2.50 | 0.50 | 1.00 | 26.56 |
| 136 | Siping | 0.80 | 5.00 | 1.50 | 1.50 | 1.00 | 1.63 | 0.50 | 2.00 | 0.50 | 1.00 | 0.74 | 2.00 | 2.84 | 2.00 | 1.00 | 0.50 | 2.00 | 26.50 |
| 137 | Qiqihar | 0.80 | 1.60 | 1.50 | 1.50 | 4.00 | 2.95 | 0.50 | 1.00 | 0.50 | 1.50 | 1.23 | 2.00 | 2.27 | 2.40 | 0.50 | 1.00 | 1.00 | 26.25 |
| 138 | Wuzhong | 0.66 | 3.60 | 1.00 | 3.50 | 3.50 | 0.00 | 3.50 | 0.50 | 0.50 | 1.00 | 0.74 | 0.50 | 2.46 | 1.70 | 1.50 | 0.50 | 1.00 | 26.16 |
| 139 | Mudanjiang | 0.85 | 2.40 | 2.00 | 4.00 | 1.00 | 0.00 | 0.50 | 2.00 | 0.50 | 1.50 | 1.36 | 0.50 | 3.02 | 1.80 | 1.00 | 0.50 | 3.00 | 25.92 |
| 140 | Yulin | 0.10 | 3.00 | 2.00 | 2.00 | 1.50 | 1.83 | 0.50 | 2.00 | 0.50 | 1.00 | 0.82 | 2.00 | 3.05 | 2.00 | 1.00 | 1.10 | 1.50 | 25.89 |
| 141 | Jinchang | 1.30 | 3.60 | 1.10 | 2.00 | 4.00 | 1.78 | 0.50 | 2.00 | 0.50 | 1.00 | 0.86 | 0.50 | 2.01 | 1.50 | 1.00 | 0.50 | 1.00 | 25.14 |
| 142 | Nanchong | 0.77 | 3.70 | 2.60 | 2.00 | 4.00 | 2.38 | 0.50 | 0.50 | 0.50 | 1.00 | 0.74 | 0.50 | 1.34 | 1.00 | 0.20 | 1.00 | 2.00 | 24.73 |
| 143 | Zhongwei | 0.61 | 1.00 | 1.50 | 2.50 | 3.50 | 1.48 | 4.00 | 2.00 | 0.50 | 1.00 | 0.68 | 0.50 | 2.25 | 1.50 | 0.20 | 0.50 | 1.00 | 24.72 |
| 144 | Ya'an | 0.10 | 0.20 | 2.60 | 5.00 | 4.50 | 1.98 | 0.50 | 0.50 | 0.50 | 1.00 | 0.83 | 0.50 | 2.83 | 1.00 | 0.20 | 0.50 | 1.00 | 23.73 |
| 145 | Luzhou | 0.96 | 5.50 | 1.10 | 2.50 | 2.00 | 0.00 | 0.50 | 0.50 | 0.50 | 1.00 | 0.78 | 0.50 | 1.73 | 1.60 | 1.00 | 1.00 | 2.50 | 23.68 |
| 146 | Qinzhou | 0.91 | 3.00 | 0.50 | 3.00 | 1.00 | 1.75 | 0.50 | 2.00 | 0.60 | 2.00 | 0.74 | 0.50 | 1.49 | 2.50 | 1.00 | 1.70 | 0.00 | 23.20 |
| 147 | Haidong | 0.30 | 3.90 | 1.50 | 2.50 | 1.00 | 1.05 | 0.50 | 0.50 | 0.50 | 1.00 | 0.72 | 0.50 | 2.45 | 2.30 | 1.50 | 1.00 | 0.50 | 21.72 |
| 148 | Anyang | 0.10 | 2.80 | 1.30 | 1.00 | 4.00 | 0.30 | 1.00 | 0.50 | 0.60 | 1.00 | 0.82 | 0.50 | 0.66 | 1.50 | 2.50 | 0.50 | 2.00 | 21.07 |
| 149 | Tonghua | 1.11 | 1.60 | 1.00 | 2.00 | 2.00 | 1.10 | 0.50 | 1.00 | 0.50 | 1.50 | 0.81 | 0.50 | 2.83 | 1.60 | 0.50 | 0.50 | 2.00 | 21.05 |
| 150 | Shuozhou | 1.20 | 1.50 | 0.50 | 2.00 | 4.00 | 0.00 | 0.50 | 0.50 | 0.50 | 1.00 | 0.77 | 2.50 | 2.77 | 1.50 | 0.50 | 0.50 | 0.50 | 20.73 |
| 151 | Luohe | 0.57 | 5.00 | 0.80 | 2.00 | 1.00 | 0.10 | 1.00 | 0.50 | 0.50 | 1.00 | 0.86 | 0.50 | 0.61 | 1.60 | 1.00 | 0.50 | 1.50 | 19.04 |

level of livelihood and open data service level. Smart economy includes urban innovation and start-up level, economic output energy consumption level, and internet industry development level. Smart population includes IT service industry workers and public life network level. Finally, security system includes planning and development, public information training and performance assessment.

- Assessment Results

Through a comprehensive assessment of 151 cities nationwide, we can obtain an overview of the development of smart cities in China. The 2015 average assessment score was 40.05 out of 105 points. The city with the highest score was Wuxi, which scored 80.20, compared with the lowest scored city Luohe, which scored only 19.04 points. In summary, the level of development of smart cities is uneven, with a significant discrete trend.

## Appendix 3: Review of Important Events of Smart Cities in China

- Smart city for the first time mentioned in the government administration report [28]
- Promoting the innovation and development of cloud computing, and fostering new forms of information industry [29]
- WIT120 – cloud medical service [30]
- Ministry of housing and urban announced the third batch of national smart city pilot list [31]
- Smart transportation – the one-card project [32]
- Smart manufacturing – Made in China 2025 strategic plan [33]
- China will take actions in software and big data industry in the thirteenth five-year development plan [34]
- Encouraging start-up companies and promoting innovation [35]
- Using big data to improve market services and supervision [36]
- "Internet+" has become a strong driving force for the development of smart cities [37]
- National Standards Committee plans to create 41 smart city international standards [38]
- Guidance on internet finance launched
- Ministry of industry launched smart city special talent certification program [39]

- The proposal on the development of national economic and social development during the thirteenth five-year plan by CPC Central Committee issued [40]

## References

[1] Beijing Municipal Commission of Economy and Information Technology. Notice on the Issuance of Beijing "Xiangyun Project"Action Plan, 2010.

[2] http://www.bjeit.gov.cn/zcjd/zcwj/113201.htm.2015.07.30

[3] Tai Chi government cloud to help Beijing to upgrade the wisdom.

[4] http://www.cnbp.net/case/detail/12351.2016.07.11

[5] Chen Jun, Li Qi, "Research on the General Framework of Beijing's Governmental Informatization," *Science of Surveying and Mapping*, vol. 08 2014, pp. 53–57.

[6] Dou Haibin, "Research on the Problems and Countermeasures in the Construction of Wisdom Nanjing," Nanjing University of Science and Technology, 2015.

[7] LI Deren, YAO Yuan, SHAO Zhenfeng, "Big Data in Smart City," *Geomatics and Information Science of Wuhan University*, vol. 39, 2014, pp. 631–640.

[8] Lionel M. NI, Zhang Qian, TAN HaoYu, LUO WuMan, "Smart healthcare:from IoT to cloud computing," *Scientia Sinica (Informationis)*, vol. 4, 2013, pp. 515–528.

[9] HKSAR Government, 2007, The 2007–08 Policy Address. http://www.policyaddress.gov.hk/07-08/eng/policy.html

[10] Sun Wende, Shen Fenggui, Zhang Weizhong, "The Present Situation and the Countermeasures of Intelligent Medical Construction in Hangzhou," *Modern City*, vol. 04, 2013, pp. 34–37.

[11] Ni Mingxuan, Zhang Qian Tan Haoyu,Luo Wuman,Yang Xiaoxi, "Smart Medical-From Things to Cloud Computing," *Science China Information Sciences*, vol. 04, 2013, pp. 515–528.

[12] http://www.h3c.com.cn/Solution/Smart_City/Smart_Community/

[13] Chen Guilong, "Biyun community: the rise of smart community," *Information of China Construction*, vol. 16, 2015, pp. 62–65.

[14] Yan Binbin, Experience and Revelation of Intelligent City Construction – Taking Taoyuan Village in Taiwan as an Example," *Contemporary Economics*, vol. 11, 2013, pp. 41–43.

[15] http://www.i-taoyuan.com/

[16] https://ntl.bts.gov/lib/59000/59200/59263/download1.pdf

[17] Palmisano, Samuel J, "A smarter planet: the next leadership agenda," IBM, 008.

[18] http://www.networkworld.com/article/2850874/big-data-business-intelligence/how-ups-uses-analytics-to-drive-down-costs-and-no-it-doesn-t-call-it-big-data.html

[19] http://www.btic.org.cn/xxzx/

[20] Xu Yanqin, "Analysis on Innovative Marketing Strategy of O2O – Analysis Based on Didi Chuxing," *Modern Business*, vol. 36, 2015, pp. 33–35

[21] Cheng Jicheng, *Smart City-Theory, Method and Application*, Beijing, Science Press, 2003.

[22] Yang Zhenghong, *Big Data, the Lnternet of Things and Cloud Computing*, Tsinghua University Press, 2014.

[23] http://english.gov.cn/12thFiveYearPlan/

[24] http://en.ndrc.gov.cn/newsrelease/201503/t20150330_669367.html

[25] http://www.besticity.com/

[26] Du, Yingsheng, and Youchun Tang. "Study on the Development of O2O E-commerce Platform of China from the Perspective of Offline Service Quality," *International Journal of Business and Social Science*, vol. 05, 2014, pp. 308–313.

[27] Wan, Jiafu, "Software-defined industrial Internet of Things in the context of Industry 4.0," *IEEE Sensors Journal*, vol. 16.20, 2016, pp. 7373–7380.

[28] http://www.china.org.cn/chinese/2015-03/17/content_35077119.htm

[29] http://english.gov.cn/policies/latest_releases/2015/01/30/content_28147 5047556064.htm

[30] http://english.gov.cn/policies/latest_releases/2015/03/30/content_28147 5080268064.htm

[31] http://www.gov.cn/gzdt/2013-08/05/content_2461584.htm

[32] http://www.moc.gov.cn/zfxxgk/bnssj/dlyss/201505/t20150505_18122 74.html

[33] http://english.gov.cn/policies/latest_releases/2015/05/19/content_28147 5110703534.htm

[34] http://www.miit.gov.cn/n1146290/n4388791/c5465401/content.html

[35] http://www.gov.cn/zhengce/content/2015-06/16/content_9855.htm

[36] http://english.gov.cn/policies/latest_releases/2015/07/01/content_28147 5138273106.htm

[37] http://www.cnnic.net.cn/hlwfzyj/hlwxzbg/201601/P020160122469130
     059846.pdf
[38] http://www.gov.cn/xinwen/2015-07/18/content_2899360.htm
[39] http://www.ntcsc.org.cn/
[40] http://www.gov.cn/xinwen/2015-11/03/content_5004093.htm

## Biographies

**Wei-Dong Hu** received the Ph.D. degree in electromagnetic field and microwave technology from the Beijing Key Laboratory of Millimeter Wave and Terahertz Technology at the Beijing Institute of Technology, China, in 2004. He is currently an Associate Professor at the Beijing Key Laboratory of Millimeter Wave and Terahertz Technology, School of Information and Electronics, Beijing Institute of Technology. His research interests include electromagnetic, terahertz (THz) technology and terahertz remote sensing.

**Jianping An** received the B.S. degree from PLA Information Engineering University, Zhengzhou, China, in 1987, and the M.Sc. and Ph.D. degrees from the Beijing Institute of Technology, in 1992 and 1996, respectively, all in communication engineering. He joined the School of Information and Electronics, Beijing Institute of Technology in 1996, where he is currently a Full Professor. He is currently the Dean of the School of Information and Electronics with the Beijing Institute of Technology, Beijing. His research interests include the field of cognitive radio and signal processing in wireless communications.

# 4

# Key Technologies and Applications for Smart Cities in China

Weidong Hu, Jianping An, Shi Chen and Xin Lv

School of Information and Electronics, Beijing Institute of Technology

## Abstract

With the advent and the rapid development of the next generation of the Internet, Internet of things, cloud computing, big data analysis and other information technology, IBM proposed the slogan of constructing "smart city" in 2008. When more and more serious urban diseases occur such as population expansion, resource shortage, serious environmental pollution, traffic congestion and increasing public safety hazards, etc.) it will lead to a helpless feeling for everyone while smart city ideas which are based on the new generation of information technology may bring new urban life expectation for the governments and people. It is with this expectation of the city's sustainable and healthy development that the smart city ideas eventually are accepted by the governments and the people. The smart city construction booms in the world.

Smart city is a large-scale joint system consisting of multiple intelligent systems from different industries. It can facilitate enterprises, government departments, financial institutions, telecommunication providers, and public sector organizations to provide better services to residents. Simply speaking, smart city is the integration of the latest information technology such as Internet of Things (IoT), cloud computing and big data technologies, as well as various network platforms. In this chapter, we first introduce the concepts and characteristics of smart city, and then elaborate that constructing smart cities in China is a major strategy in economic and developmental transformation, and an important step towards an innovation-oriented country. Next, the chapter describes with full details the sub-systems and key technologies of the integrated smart city system.

*Breakthroughs in Smart City Implementation,* 87–124.

**Keywords:** Smart City, Internet of Things, Cloud Computing, Big Data Analysis.

## 4.1 Introduction

Introducing and integrating Smart City concepts have led to a breakthrough in world-wide scientific and technological revolution after industrialization, electrification and informatization. Using intelligent technologies to build smart cities is the global trend nowadays. With the rise of wide-spread technologies such as IoT (Internet of Things), cloud computing, mobile internet, and big data analysis, the concept of smart city is getting supreme attention in all major countries. From the establishment of People's Republic of China in 1949 to the reform and opening up, and then to the sustainable development in the 21st century, China's urbanization process continues to deepen. The number of cities increases significantly, accompanied with fast growing urban population and expanding city scales. Cities have become the driving force of China's GDP growth. Constructing smart cities will become the key breakthrough in China's urban management and will lead to new urban outlooks in the future. Scientific research and commercial applications in fields such as smart environment, E-government, smart tourism, smart health care, smart transportation, wisdom education are becoming top priority topics [1–4]. At the same time, operation service providers, solution providers, terminal manufacturers and application developers are establishing their positions in the smart city industrial chain. Construction of smart cities has spread out all over China. Governments at all levels have gradually determined the construction scale. The construction of smart cities has provided a platform for new technologies with important opportunities to experiment. In return, these technologies have also promoted the further development of smart cities and technology integration.

## 4.2 Definition of Smart Cities

Smart cities utilize information and communication technologies to sense, analyse and integrate key information in urban operation core systems, so as to make intelligent responses to various needs. These needs include people's livelihood, environmental protection, public safety, urban services, and industrial and commercial activities [5]. In essence, it is applying advanced information technology in urban management and operation, in order to

**Figure 4.1** Smart city Nansha in Guangzhou.

make a better life for urban residents and to promote the harmonious and sustainable development of cities. IBM brought up the concepts of Smart Earth and Smart City in 2008 for the first time [6] and thereafter Smart City has become the vision and target of government administration in many countries. According to a report on European smart cities published by the European Commission [7], a smart city can be characterized by six elements: smart economy, smart people, smart governance, smart mobility, smart environment and smart living. In short, the nature of smart city is high integration of urban services which require advanced urban informatization. As the new the engine of economic restructuring, industrial upgrading, and city levelling, smart city is regarded as an integrated development strategy for improving the well-being of people, increasing the competitiveness of enterprises, and eventually achieving sustainable development of cities.

## 4.3 Characters of Smart Cities

In order to make a quantitative characterization of smart cities we use a formulation which is commonly used in describing operational research problems.

### 4.3.1 Smart City Described as Operational Research Problem [8]

Assume that matrix $A_0$ is the resource allocation of a city at time 0, and matrix $S_0$ is the set of exogenous states this city faces. Assume that given these exogenous states, the resource allocation at time 0 is optimal; for example, a newly built city.

Therefore,

$$A_0 = \arg\{\max_A V(S_0)\} \tag{4.1}$$

where $V$ is the welfare function of the city. $V$ can be interpreted as the sum of maximum values from welfare function over all periods in the future.

At time 1, things change so that the states become matrix $S_1$. Decision makers have to decide whether to change the current resource allocation or not, and if so, how to change?

Specifically, the cost function $\phi$ is defined by maximizing the function

$$\max_A \{V(S_1) - \phi(A - A_0)\} \tag{4.2}$$

This is an optimization problem with adjustable costs, where the cost function satisfies:

$$\phi(0) = 0, \phi(X) > 0, \quad \forall X \neq 0 \tag{4.3}$$

It is easy to understand that if adjusting is too expensive, then the optimal solution might be status quo because the costs of adjusting will exceed the benefits of adjusting.

If there are such technologies available that can decrease the costs of adjusting, then the possibility of implementation of optimal resource reallocation will increase correspondingly. We cannot say that every single such technology will lead to an optimum resource reallocation because one single technology might not reduce the cost of adjusting much. However, as the city applies enough new technologies, the cost reduction will accumulate so that finally it becomes feasible for the city to allocate its resources according to variable states and to increase welfare of its people. Mathematically, the function

$$A_1 = \arg\{\max_A [V(S_1) - \phi(A - A_0)]\} \& V(S_1) - \phi(A_1 - A_0) > V(S_0) \tag{4.4}$$

describes an optimization problem with adjustable cost function and can help us to understand two things:

a. Why many things are apparently unreasonable or outdated but have not yet been optimized?

- Equation 4.4 learns: the costs to adjust are too high.

b. Why is smart city a science rather than a pure concept?

- Equation 4.4 learns: the technologies and the scientific approach behind the concept can bring opportunities to reduce the cost to adjust.

However, merely knowing this kind of optimization models is not sufficient to solve the problems. Solving problems requires cooperation from professionals of at least three fields:

- Professionals from industry, who can calculate the precise costs of applying technologies and describe the situations where technologies have been applied;
- Juridical professionals, who can identify the risks associated with information security and estimate the costs to eliminate them. They also make legislative proposals targeted at eliminating risks;
- Economists, who can perform cost accounting to increase benefit with information provided by the first two kinds of professionals.

It appears that technology is the biggest obstacle. The last two kinds of professionals have to rely on a certain technology, and based on this specific technology they can analyse costs, benefits and risks.

From the core characters of smart cities as just described, we must fully understand "1C3W1H", i.e., 1 times C or the context of smart cities, 3 times W or what are smart cities, when are smart cities built, who should smart cities, and 1 times H or how to build smart cities. 1C3W1H will be detailed in the paragraphs from Section 4.3.2 to 4.3.6, but first we elucidate that smart cities should be people-oriented.

## 4.3.2 People-Oriented: The Core Character of Smart Cities

The essence of any technology is to make people more satisfied, to make the environment more liveable, to create more opportunities for developments, and to make communications smoother among people. The construction of smart cities should start from its importance for urban residents and should emphasize people's experience. The new value of constructing smart cities is to enhance the human care, to promote the culture and to increase the public's sense of happiness.

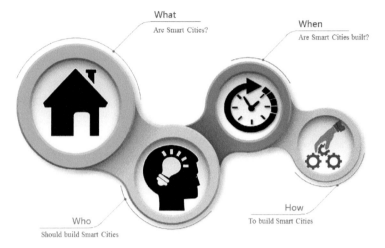

**Figure 4.2**   "1C3W1H".

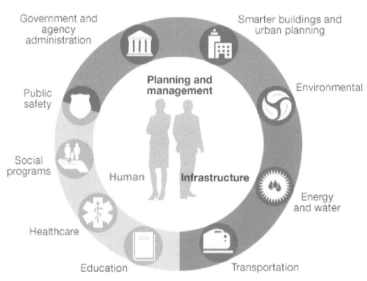

**Figure 4.3**   Smart city initiative (IBM, 2008).

### 4.3.3 Context of Smart Cities

With the rapid development of internet technologies, the era of big data has arrived. The technical conditions for constructing smart cities are ready. Although the birth of internet is only forty years ago, it has greatly changed the way of living but also the way of producing goods. Internet is a new dimension that is different from the space of real life. It is our second space and a virtual space. Based on hardware and operation rules, the core elements of internet space are IDs, data and links, while the core elements of real life are people, things and objects. Every element in the two spaces can overlap and interact. Real life maps onto internet and creates enormous data. Mining and using data from internet can also react on real life. As a result, cities become smart. The interaction between the two spaces has built the technical foundation of smart cities.

### 4.3.4 What Are Smart Cities?

Smart city is a goal of urban development. In the information age, it is also the imagination of modern cities about the future – more suitable for living and working, more lively, more attractive and more competitive. Smart city is also a development scheme. It requires switching from the traditional extensive economy which relies heavily on inputs. By applying new technologies and involving multiple parties, we can achieve optimal resource allocation and increased production efficiency. In short, more value can be created with fewer resources.

### 4.3.5 When Are Smart Cities Built?

Smart city is a development process that includes constant innovation and continuous development and improvement, given the city's current states. With continuous breakthroughs of technology and improvements of knowledge, the content of smart city will become richer. Therefore, there is no ultimate state of completion.

### 4.3.6 Who Should Build Smart Cities?

Smart city is an open platform that welcomes multiple participants. Governments, enterprises, research institutes, and residents all should participate in the construction of smart cities. Different parties are assigned with different tasks. For example, governments launch policies and regulations, enterprises produce goods and provide services, research institutes release standards and make recommendations, and residents participate throughout the entire process. They are the users and reviewers.

## 4.3.7 How to Build Smart Cities?

To construct smart cities, it is necessary to take a top-down method in planning and designing. The first step is strategic planning by which we can grasp the overall direction of constructing smart cities. This is followed by top-level design and action plans. Gradually we can implement the strategy and carry out sustainable operations. Apparently, this process needs ongoing actions to ensure that the strategy can continue maintaining its advanced and leading position.

An information resource centre is the foundation of constructing smart cities. Urban operations generate a large amount of data. Organizing and analysing the facts carried by data can form valuable information contents and information products. Moreover, it can inspire service innovation to achieve economic benefits, social benefits and environmental benefits, and finally advance cities toward a less-carbon, more efficient and convenient future. Therefore, in order to promote the effective use of data, it is necessary to establish an information resource centre in advance as the base of construction. The information resource centre will collect data from all levels of urban operation and extract valuable information for developers from various fields, acting as the engine of innovation and development.

An object-oriented integrated service platform is the key to constructing smart cities. It is also the interactive connection between reality and internet. The basic principles of planning and constructing smart cities are "future-oriented", "problem-targeted", and "object-oriented". Future-oriented and problem-targeted are more for the strategic planning stage. Object-oriented means that during the construction process, the integrated service platform has to orient itself to three primary objects, i.e., residents, enterprises and governments. In other words, the service platform includes platform for residents, for enterprises, and for urban operation and management. Constructing smart cities shall use the three platforms as starting point and focus. This will bring together a number of object-oriented services to provide a full range of multi-channel seamless service experience, to connect stakeholders of smart cities, and to promote the sustainable development of cities with information technologies.

Smart cities and ecosystems share many features in common: diversified objects, ubiquity, energy input and flow, self-adaptability, and so force. Among them, there are three most important features, explained next.

Diverse elements, which constitute the basis of smart cities. Like the fundamental character of an ecosystem is biodiversity; the elements that constitute smart cities are also diverse. Individuals, organizations made from individuals, infrastructure and the natural environment; they together form the cities. Each of the elements has different attributes, capabilities and characteristics. The basic element of cities is the individual. Each individual has a different identity and label: he can be the father of a family, a staff of a company, a customer in a store, or a pedestrian in streets and so on. Different identities correspond to different behavioural traits, different daily needs and different social patterns. Combinations of individuals form different organizational units, including families, companies, schools and government agencies. There are also a variety of infrastructures in cities. They are human transformation of the natural environment. These infrastructures serve the basic needs of survival. As civilization develops, various types of infrastructure go beyond functional needs. Aesthetic needs are also satisfied. Natural environment is the most fundamental element of cities. It is not only being transformed passively, but also adaptive to changes actively.

Dynamic equilibrium aims that information flow between elements allows smart cities to achieve dynamic equilibrium. In ecosystems, energy flows freely along the food chain and thus keeps the population structure relatively stable. For smart cities, information resources are the flowing energy. The process of generating, aggregating, refining and using information is also the process of gathering and consuming energy. Elements in cities are both producers and consumers of information. Taking traffic as an example, pedestrians' individual behaviour data are collected and submitted to urban traffic index data as a whole. Once analysed by the intelligent traffic system, control methods such as signals are used. These methods react on pedestrians so as to reduce traffic pressure and relieve congestion. In this process, the flow and refinement of information has created extra value, which is also consumed. The input and output of energy and matter in ecosystems make the system stable. Similarly, the flow of information in smart cities can create inestimable value to sustain elements at equilibrium level. The expression of such equilibrium is diverse. To name a few, the population structure stays relatively stable and healthy; the natural environment is not destroyed due to excessive human activities; buildings are strong and energy-saving; traffic is not over-congested.

Open and developing is based on the principle that an open attitude to the outside world makes that smart cities continue to develop and become more perfect. Ecosystems are open to energy and can internalize energy for their own development to achieve continuous evolution. Our present life greatly relies on data, knowledge, and intelligence. Life, production organizations, as well as social organizations have experienced tremendous change. Innovation has become the theme of the period. Knowledge has become the most important factor of production and also of the product itself. Internet has become a significant media where information and commodity can exchange. The core character of smart cities is openness. It does not mean a simple investment, but a friendly environment that can cultivate knowledge and innovation so that cities become more suitable for innovative talents to develop themselves, and innovative enterprises to settle and operate. The future of smart cities will rely on an open and innovative economy to achieve more intensive, more efficient and more sustainable growth.

## 4.4 Importance of Constructing Smart Cities

The ecosystem of a city is mainly composed of five major functional systems: resources and environment, infrastructure, economy and industry, municipal governance, and social and livelihood. The past twenty years has witnessed rapid urban development. The five functional systems are also facing unprecedented challenges (Figure 4.4, Table 4.1) [9]. Further urbanization requires improving the scientific level of urban construction and management.

**Figure 4.4** Ecosystem of city.

**Table 4.1** Problems of urban functional systems

| Challenges | Phenomena |
| --- | --- |
| Urban infrastructure to be improved | Heavily congested urban traffic |
| Governance efficiency to be improved | Inefficient government |
| | Difficult to use public service |
| Livelihood to be improved | Unsatisfying social security, health, education and other service |
| Economy and industry to be developed | Structure to be optimized Traditional resource- and labor-intensive to be transformed |
| Resources and environment to be improved | Pollution and energy over-consumption Unsustainable development |

Improving the efficiency of urban construction and management is an urgent need for the stability of countries and the happiness of residents. Smart city can provide an effective solution to the problems that cities are facing. The idea of constructing smart cities is supported by the scientific outlook on urban development, and it utilizes the new generation of information technology comprehensively. Infrastructure such as broadband multimedia information networks and cloud computing platforms will be built to integrate smart cities' information resources on a unified online platform, so as to provide convenient public services to their residents. The platform will also assist in public administration to provide efficient and competitive solutions to various demands, including municipal monitoring, smart transportation, electronic health care, smart tourism, public security and so on. Moreover, the platform helps enterprises to increase efficiency and thus to enhance industrial capacity. Ultimately, it makes smart city invincible in competition in the information age.

## 4.5 Main Technologies of Smart Cities

### 4.5.1 Internet of Things (IoT)

#### 4.5.1.1 Introduction to IoT

The concept of IoT was brought up in 1999 [10]. It was initially defined as "sensors and actuators embedded in physical objects are linked through wired and wireless networks". At the World Summit on the Information Society held in Tunis on 17 November 2005, the International Telecommunication Union (ITU) released the *ITU Internet Report 2005: Internet of Things* [10], formally proposing the "Internet of Things" concept. Internet of things is defined

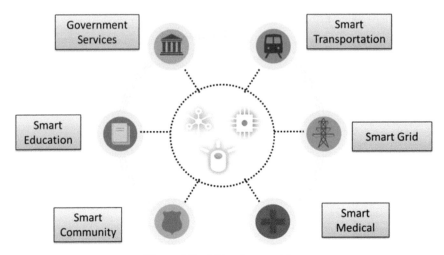

**Figure 4.5**   IoT and smart cities.

as a network that through radio frequency identification, infrared sensors, global positioning system (GPS), laser scanner and other information sensing equipment, and according to agreement, connects any item with internet and exchanges information and communication to intelligently identifying, locating, tracking, monitoring and managing. In particular, the sensor is embedded and equipped to the power grid, railways, bridges, tunnels, roads, buildings, water systems, dams, oil and gas pipelines and other objects, and is generally connected to form the IoT.

As early as 1999, China has put forward relevant concepts. It was known as the sensor network and the Chinese Academy of Sciences launched research and development on it. Given the background that the country is vigorously promoting industrialization and information technology integration, IoT is a realistic breakthrough in the process of industrialization and information technology integration. IoT can connect people to people, people to machines, and machines to machines, therefore give full play to the advantages of people and machines.

Initially the core application of IoT was identifying objects with electronic tags (RFID). As technology developed, sensors and chips have been embedded into objects to form a machine to machine (M2M) network which becomes a part of IoT. From IoT's data acquisition function point of view, earth observation technology and remote sensing technology can also be integrated into IoT.

The application of IoT has changed the information interaction pattern where human-computer interaction dominates in digital cities. Now it has come to the stage where sensors and chips are ubiquitous, with real-time information acquisition and intelligent control. It brings intelligent environment opportunities to develop.

The development of earth observation technology, especially high resolution satellite imaging, unmanned aerial vehicle (UAV) remote sensing, light detection and ranging (LIDAR) and continuous operation reference station (CORS), as well as the recent light field camera technology, will make it easier and more efficient to acquire geospatial data (including 3D spatial data), and therefore provide more data sources for the digital urban geospatial framework and applications in various fields.

The development of IoT has fundamentally improved the information acquisition capability from macro to micro levels. It has also increased the macro-and micro-control capability of cities. Finally, it can enhance the intelligence of the human-centred environment.

IoT is the derivative and application of communication networks and internet. It employs sensing technologies and intelligent devices to sense and identify the physical world, and then transmits its data through internet. After computing, processing and information mining, IoT connects people to machines to exchange information. The goal is to achieve real-time control of the physical world and to allow for precise management and scientific decision-making.

### 4.5.1.2 Characters of IoT

The structure of IoT can be divided into three layers: the perception layer, the network layer and the application layer. The perception layer perceives, identifies and controls the physical world. The network layer transmits information. The application layer is the place where applications in various organizations are realized after calculating and processing information.

The key technologies of IoT include sensing and RFID integration technology, identification and environment detecting technology, IoT nodes and gateway technology, IoT communication and frequency tube technology, IoT accessing and networking technology, IoT software and algorithms, IoT interaction and control, IoT computing and services, and so on.

### 4.5.1.3 IoT and smart cities

Applications of IoT can be found in all aspects of the smart cities, and provide comprehensive support for the perception and control of smart city information system. The applications are manifold, some of which in different areas are listed below:

- Industry. Examples are production process control, supply chain management, energy consumption control.
- Agriculture. Examples are fine irrigation of crops, growth monitoring, circulation of agricultural products.
- Business. Examples are vending machines, Point of sales (POS) terminals.
- Financial services. Examples are "Golden Card Project", the second generation ID card.
- Transportation. Examples are traffic flow monitoring, traffic signal control, electronic tolls, navigation systems, vehicle condition diagnosis.
- State power. Examples are intelligent substation, intelligent electricity, and automatic distribution.
- Medical and health. Examples are remote diagnosis, medical waste monitoring.
- Education. Examples are book information push and distance education.
- Home-use facilities. Examples are access control, security, remote control of electrical equipment.
- Environment. Examples are harmful substances detecting, climate monitoring, environment monitoring.

At present the development of IoT experiences a phenomenon known as information isolation. Precisely, applications are in isolated geographical regions and in various industries. It is difficult to form a city-wide integrated collaboration platform. In addition to the socio-economic difficulties, the main reasons are the low degree of standardization, the lack of a unified middleware interface, and the integration pressure resulted from massive information storage and computing from various departments. Luckily, cloud computing technology can provide a good solution to the last one.

## 4.5.2 Big Data Technology

### 4.5.2.1 Introduction to big data

A smart city generates massive amount of data every day. These data are characterized by large volume, diversification, rapidity and low value density. In other words, only the data with these characteristics can be considered as big data. It is difficult for traditional data analysing technologies with conventional software tools to capture, manage and process such volume of structured data, semi-structured data, and especially unstructured data (pictures, video, sound and other documents) in a certain period of time [11]. Big data analysis is the forefront technology of data analysis. It can extract useful and valuable information from the above-mentioned types of data. Key technologies for integrating, processing, managing and analysing big data include business intelligence, cloud computing, data warehousing, data marts, distributed systems, metadata, non-relational databases, relational databases, stream processing, visualization technologies and so on.

Generally, big data can be characterized with four Vs, i.e., volume, velocity, variety and value.

Volume refers to the large capacity, from Tera Byte (TB) level to Peta Byte (PB) level. Velocity means that big data grows fast and hence requires fast processing speed. Variety refers to the rich data types, including structured data and unstructured data. Value refers to low-density of data value. In other words, the percentage of data that carry value is small. Big data has simpler but more efficient algorithms compared with small data's intensive algorithms. Samples in the context of big data are no longer random samples, but the whole population. Moreover, big data analysis is not an accurate but rather hybrid method. It investigates correlation instead of causal relationships.

IoT and internet also raise problems related to storing, managing, processing, integrating and mining the multi-source massive data. Traditional relational database management system can no longer fully adapt to the management and computing demand. Currently, key technologies that are used to integrate, process, manage and analyse big data include BigTable, business Intelligence, cloud computing, Cassandra, data warehouse, data mart, distributed systems, Dynamo, GFS, Hadoop, HBase, MapReduce, Mashup, metadata, non-relational database, relational database, R, structured data, unstructured data, semi-structured data, SQL, stream processing, visualization technology and etc [12].

Specifically, key technologies for big data analysis include A/B testing [13], association rules mining, classification, data clustering, crowdsourcing, data fusion and integration, data mining, integrated learning, genetic algorithms, machine learning, natural language processing, neural network, neural analysis, optimization, pattern recognition, prediction model, regression, emotion analysis, signal processing, spatial analysis, statistics, supervised learning, unsupervised learning, simulation, time series analysis, time series forecasting model, visualization technology and etc.

### 4.5.2.2 Characters of big data analysis

Data from smart cities have the following traits:

- Diverse sources of data

  In order to achieve information sharing and intelligent response between urban systems, public supporting platforms of smart cities have to gather various kinds of data. These data come from various industrial systems and urban basic information databases. Transportation, municipality, environmental protection and other industrial systems generate data of the city's operating status, while urban basic information databases provide data on demography, legal persons, geography and economy. Smart city public supporting platforms will have to successfully dock with the various heterogeneous systems.

- Diverse types of data

  The data types in smart cities include structured data, semi-structured data, and unstructured data. Structured data can be expressed with a two-dimensional table structure. extensible markup language (XML), hypertext mark-up language (HTML) and other mark-up languages have structure and labels that can self-describe. They belong to semi-structured data. Unstructured data are those that do not have pre-defined data model, or are not suitable for relational database to save information. Usually they are stored in the form of documents. Take temperature and humidity as example. The gas concentration sensor creates structured data that can be directly stored in the relational database, while the monitor creates unstructured data. Videos are processed with intelligent algorithms and tagged. After structural semantics processing, they become semi-structured data.

- Tremendous data size

  The increase in the size of cities also lead to a dramatic increase in the volume of data. In 2007, the number of urban residents reached 3.3 billion. By 2050, the number is projected to exceed 70% of world's population, i.e., there will be 6.4 billion urban dwellers.

Informatization has gradually digitized a large number of previously non-digital information, leading to exponentially growing data size. At the same time, as urban functional infrastructure gradually realizes object-to-object, a large number of perception data are generated. A medium-sized city needs to deploy 300 to 400 thousand cameras and totals 500TB data in a month.

Because traditional information processing technologies are unable to deal with multi-source, heterogeneous and massive data, big data technologies have inherent advantages in processing such data.

### 4.5.2.3 Big data technology and smart cities

- Big data provide strong decision support for government management.

In urban planning, data mining of natural information and social information can provide strong decision support for urban planning, and strengthen the city management service in a scientific and forward-looking way. Natural information includes geographic and meteorological information, and social information includes economic, cultural and demographic information and so on.

In traffic management, data mining of real-time traffic information can effectively alleviate traffic congestion and quickly respond to unexpected conditions, thus provide a scientific basis for decision-making in urban traffic operation.

From public opinion monitoring we know that keyword searching and semantic intelligence analysis can improve the timeliness and comprehensiveness. Big data also help to get a comprehensive understanding of social conditions and public opinions, to facilitate public services, to cope emergencies, and to combat crimes.

In public security, mining of big data helps to detect man-made or natural disasters and terrorist events timely and increase capabilities in emergency response and security.

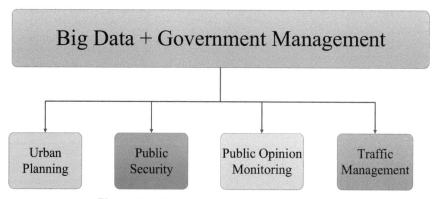

**Figure 4.6** Big data for government management.

- Big data can improve the livelihood of residents in cities

Smart applications such as smart transportation, smart health care, smart home, intelligent security, etc [11] are closely related to our daily life. These applications can greatly expand people's living space, and transform urban lifestyle into intelligent lifestyle in the age of big data. Big data is the basis for people to enjoy an intelligent lifestyle at present and in the future. Thanks to the applications of big data service, traditional "ordinary" lifestyle has been changing. Information will become more ubiquitous, and life will become multi-dimensional and more plenitude.

- Big data analysis determines company's core competences

Whoever controls the relevant data also can dominate the market and enjoy high return on investment. In the past, many enterprises would analyse their operation and development based on simple business information, which lacks customer demands, business processes, marketing, market competition and other aspects of in-depth analysis.

If decision-makers rely solely on the current situation of their business and their subjective experience to make predictions on markets, it will lead to risks and even failures in strategic planning and decision-making. In the age of big data, enterprises are able to collect and analyse massive internal and external data to extract valuable information. By data mining such information, they can estimate market demands and make decisions intelligently. Studies [12] have shown that in US companies, product and service quality will increase by 14.6% if the level for making intelligent decision is increased by 10%.

### 4.5.3 Cloud Computing

### 4.5.3.1 Introduction to cloud computing

Cloud computing is a network-based service provision model that supports heterogeneous facilities and resource flows. It provides autonomic services to customers. Cloud computing supports heterogeneous resources and heterogeneous multi-tasking systems [14]. This will result in on-demand distribution of resources, billing according to the amount of use, and consumption based on actual demand. Ultimately, it will increase the scale of resources and the degree of specialization. Unit cost per resource will decline. Innovation on online business will be inspired by cloud computing.

Smart city is a complex composed of multiple applications, various industries, and many complex systems. There are needs of information sharing and interaction between multiple application systems.

Different application systems will have to extract data collectively to synthesize and present the comprehensive results. The number of complex systems requires sufficient and powerful information processing centres to process a variety of information. In order to provide steady and safe support for the operation of large systems, we need to take into account the cloud-based network architecture when build smart city cloud computing data processing centres. Meanwhile, these centres have certain advantages which traditional data processing centres or single application systems have not. First, dynamic scalability based on demand. Platforms are built based on cloud computing, so they can add application systems dynamically. Second, high performance investment ratio. Relative to the traditional data processing centre, cloud computing data processing centres can decrease hardware investment by at least 30% [15].

### 4.5.3.2 Characters of cloud computing

• Unification and high performance of the platform layer

Using infrastructure-as-a-service (IaaS) models to integrate servers at traditional data processing centres that have different structure, different brands and different models will provide a unified operation support platform to the application system through cloud operating system scheduling. At the same time, with the help of the virtualization infrastructure of cloud computing platform, it is possible to carry out resource partition, resource allocation and resource integration effectively and to allocate computing and storage capacities based on needs to achieve optimal performance.

• Large-scale basic software and hardware management

Basic software and hardware management mainly works on monitoring and managing large-scale basic software and hardware resources. It also involves resource scheduling for the operating systems at cloud computing centres. Basic hardware resources include three major devices in the network environment: computers (servers), storage (storage devices), and networks (switches, routers, and etc.). Asset management on software and hardware resources can be carried out at the management centre so that status monitoring and performance monitoring can be achieved. Abnormal situations will trigger alarms at the management centre and prompt users to fix the problem equipment. Software and hardware management can also provide the decision basis for high-level resource scheduling through long-term analysis.

- Business and resource scheduling management

Cloud computing data processing centres are characterized by a large number of basic hardware and software resources. In other words, the scale of basic resources is tremendous. These centres can improve resource utilization rate, reduce unit cost, and enable multiple users to share resources. They can also allocate resources automatically to where they are scarce according to business load information. Business and resource scheduling centres are advanced applications of the operating systems at cloud computing data processing centres. The scheduling centres are also a necessary requirement of low-carbon and green businesses.

- Security control management

Industry and countries can have a lot of competitiveness. At the information age, data have become the core competitiveness of any industry and any country. Cloud computing has made it possible to separate computing and storage, so that a large number of users can benefit from the same basic

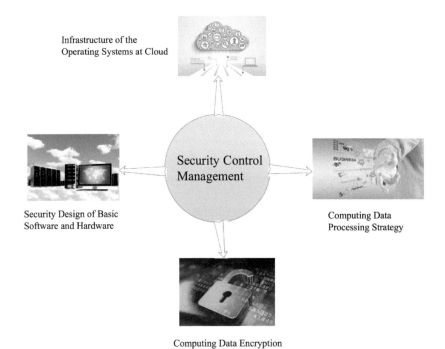

**Figure 4.7**   Security control management.

resources by sharing. However, many users sharing the same resources at the same time also raises challenges on data security.

In the cloud computing environment, the centralized management of basic resource has transferred end users' security issues to data processing centres. From a professional point of view, end users can use the security mechanism of cloud service to protect business security without spending their excessive resources and energy. But on the other hand, the cloud computing data processing centres will have to be responsible for the increasing security needs. In particular, the security of cloud computing involves the following main aspects: data access risks, data storage risks, information management risks, data isolation risks, legal investigation risks, and sustainable development and migration risks.

The security control of cloud computing data centres can be strengthened in many ways. To name a few: security design of basic software and hardware, infrastructure of the operating systems at cloud computing data processing centres, strategy, authentication, encryption and many other integrated aspects. Prevention and control will ensure the information security of cloud computing.

- Energy saving management

Building a conservation-oriented society is the basis for a sustainable economic and social development and also an important measure to protect economic security and national security. For cloud computing data centres which face large-scale basic hardware and software resources, green and energy-saving operations and maintenance of these basic resources are not only necessary for resource providers to do business, but also one of the original intention of the development of cloud computing.

At present, the general situation in industry is to procure equipment to ensure the peak needs of their various business. As a result, users tend to purchase an excessive amount of equipment. The actual situation is that the operation load is generally low, usually only about 20 percent of the full load. Particularly, at low load times the server utilization rate is even lower. Long-term low utilization has resulted in an amount of resource waste and energy loss. Cloud computing data centre can effectively improve the utilization of resources and realize multi-tenant applications by analysing historical statistics along with the business and resource scheduling and management. In a typical application, a cloud-based data centre with energy-efficient technologies can increase the resource load to 80 percent.

After removing the loss from scheduling, it can still increase the effective load of resources by a factor of two.

Currently, the number of servers in China is more than two million. If in future wide application of cloud computing energy-saving technology is to be realized, we can reduce energy consumption by about 65 percent per server location. Assume the average energy consumption of a server is 200 watts per hour, the annual saving of electricity is 16 billion kWh, equivalent to the total electricity generation at Gezhouba Power Station in 2009. In the night when the overall data center load is low, we can put the non-occupied resources into a so-called "sleep mode" so as to achieve the maximum degree of green, low-carbon energy-saving operations.

### 4.5.3.3 Cloud computing and smart cities

• Cloud computing makes cities more intelligent

Cloud computing platforms are the brains of cities thanks to their unprecedented powerful data analysis and calculation capabilities. They can fully coordinate all aspects of urban life, achieve massive data calculation and storage, improve various resource utilization rates, and save the cost of building smart cities. A smart city contains all aspects of urban life. It is a complex characterized by multi-application, multi-industry and multi-system aspects. In the construction of smart cities, a large amount of data information is accumulated. There is also need for resource sharing and information exchange between multiple application systems. All of these applications need the various data stored in cloud in order to function well. The number and complexity of the systems requires multiple powerful information processing centres to carry out a variety of information and data processing tasks. The characteristics of cloud computing can fully satisfy the requirements of smart city urban construction.

As a new computing method and service method, cloud computing is known for its massive storage capacity and variable computing power. It is provided to users in the form of services, so that users can benefit from the service in different places, at different times and on different platforms. Cloud computing's storage capacity and computing power has amplified the value and advantages of network resources, while at the same time reduced the dependence on terminal platforms. Cloud computing has steadily supported various applications of smart cities. Cloud computing has become the core of many intelligent applications. These widely-used technologies include

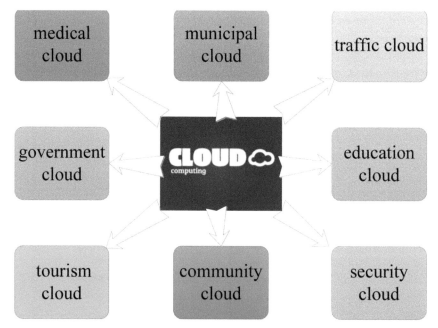

**Figure 4.8** Cloud computing applications.

e-government cloud, medical cloud, municipal cloud, traffic cloud, education cloud, security cloud, community cloud, tourism cloud, and so on. This will reform the development model of cities and greatly improve their intelligence level. The goal is to promote the healthy and rapid development of all aspects of the city.

- Cloud computing can contribute to the integration of urban information resources and promote social and economic development

The various fields of any city are interdependent and therefore shows a network relationship. The construction of smart cities requires fully understanding of information on all aspects of the city and the reticular relationships among the various fields. Smart cities will store in the cloud important information on its talent flow, material flow, information flow, resource flow, capital flow and etc. It is a more comprehensive integration of urban information resources.

Integration is the essence of smart cities. Information technology and traditional industries are integrated to produce a series of new industries in this process, such as express delivery, network operation and maintenance management, high-end site construction and so on. These new industries can indirectly benefit the development of urban traditional pillar industries. In the process of constructing smart cities that are based on cloud computing platforms, it is necessary to establish intelligent government systems as the first step. Next we can solve a series of social management problems such as transportation, medical treatment, education and livelihood by following the instructions of the intelligent government. In short, cloud computing has provided a broader space for the development of smart cities.

Smart city construction requires not only collaboration among all stakeholders, but also integration of relevant information resources. Particularly, it is necessary to emphasize the data mining, integration and reapplication of information. As an emerging computing method, cloud computing has an important function of integrating resources. This function can provide powerful support for applications to achieve information-sharing in all aspects. Because cloud computing is helpful in forecasting and decision-making, government's administrative capacity can be enhanced. For example, some cities have already developed several platforms for e-government and e-commerce. Although these platforms have accumulated a lot of valuable information resources, they are independent of each other and there is no information exchange or resource sharing. At the same time, too many repetitive system platforms make the equipment utilization rate relatively low and management costs high. The construction of the cloud computing centre can effectively integrate the hardware resources and information data of the equipment, support larger scale applications, handle larger scale data, and can dig the data more deeply. In this way, cloud computing can provide a platform for government decision-making, enterprise development and public services.

- Cloud computing can reduce the total costs by sharing available resources

An outstanding feature of cloud computing is that it enables resource sharing. Smart cities based on cloud computing can greatly improve the utilization rate of urban basic resources and effectively reduce the costs of urban infrastructure [14]. In addition, cloud computing can save energy when the load is low at night. It is possible to transfer operations to some of the actual working physical resources, and shut down or turn to an energy-saving mode the idle physical resources. Through centralized resource management, cloud computing data centre can reduce daily maintenance work and move an amount of work to professionals at backstage, thereby reducing management and maintenance costs, and improving management efficiency. To summarize,

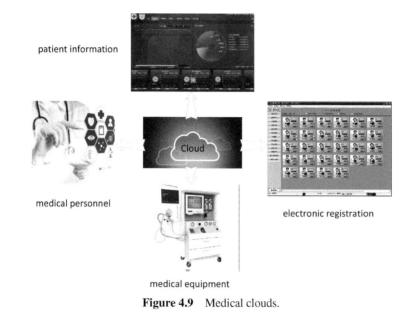

**Figure 4.9**   Medical clouds.

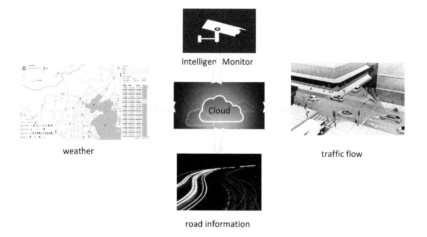

**Figure 4.10**   Traffic clouds.

resource integration and centralized management in cloud computing data centres can effectively reduce the cost of information sharing and lower the threshold of informatization so that more enterprises and government agencies can enjoy the advantages of information age and improve their efficiency.

- Intelligent cloud will support urban development by various means

Information is the basis of urban development in modern times because people demand higher quality of information services. Rather than various mixed public clouds, industrial clouds can provide information services that are richer and more professional. These specialized clouds are doing a better job in collecting industrial information and promoting the construction of smart cities. For example, medical clouds are specific for medical equipment, medical personnel, patient information, and electronic registration. Traffic clouds are specific for road information, traffic flow, weather and temperature. Commercial clouds are specific for business organizations and market information. Industrial clouds cover a number of different levels and have different sizes. They form a set of organic urban ecosystems to support smart cities and to promote their further development.

- Conserve energy to make a greener city

Too many software and hardware resource bases will result in huge energy consumption, which is contrary to the goal of a green and energy-efficient industry. It is also one of the original drivers of the development of cloud computing.

Cloud computing data centre is a common pool of resources that can be leased to multiple users, and therefore bring together a number of hardware resources for collective maintenance. In this way, energy consumption is reduced. Meanwhile, spare resources can be turned to sleep mode or simply shut down at night when load is low, so as to achieve maximum energy-saving and a greener, low-carbon environment.

## 4.5.4 Spatial Information Technology

### 4.5.4.1 Introduction to spatial information technology [15]

In 2006, Nature published a cover paper *2020 Vision*, which believed that observation network will enable large-scale real-time access to real-world data for the first time [16]. Nowadays, the space-air-ground information observation and acquisition systems have basically taken shape. The way to collect spatial information has changed from traditional manual measurement to space-borne remote sensing platforms, global positioning and navigation systems, airborne remote sensing platforms, and ground vehicle mobile measurement platforms. Acquiring and updating spatial information has become faster and faster. Positioning technologies are expanding from outdoor to

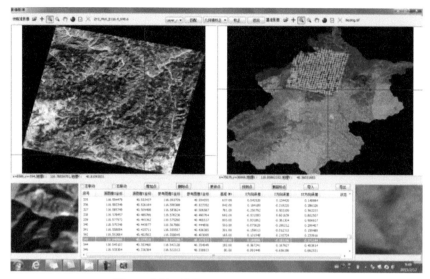

**Figure 4.11** GIS systems and smart cities.

indoor and underground. The amount of data from multi-resolution and multi-temporal observation and measurement is exponentially growing. Digital cities have the ability to monitor spatial information of different resolution, including land types, buildings, roads, municipal facilities and etc [17].

As one of the 16 major projects on the national science and technology mid- and long-term plan, the high-resolution earth observation technology research and development project was initiated successfully in 2006 [18]. By 2020, China will have launched 14 own-designed high-resolution satellites, each downloading an estimate of 2PB of data daily. This project will increase the optical and radar satellite remote sensing resolution to 0.3 m in order to meet the needs of smart cities. Now Satellite Number 3 in orbit can reach 2-3m accuracy. The next mapping high-resolution Satellite Number 7 will further improve the accuracy to 0.6 m. The commercial remote sensing satellites that will be launched this year will also be used for constructing smart cities.

By 2020, China will build an advanced observation system composed of space-air-ground observation platforms to monitor the atmosphere, land and sea. China will for the first time connect the space-air-ground network and achieve multi-sensor collaborative observation, real-time data processing, and rapid response of application services [19]. It will greatly improve the ability and quality of earth observation to provide a stable

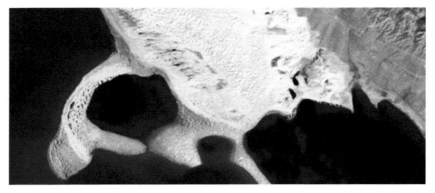

**Figure 4.12**  Satellite number 3.

operation system. Finally, the observation system will improve China's spatial data self-sufficiency rate and create a spatial information industry chain.

### 4.5.4.2 Remote sensing technologies

It is generally believed that with the development and wide application of mobile technologies, we have entered a new era of collecting and utilizing big data. In fact, we cannot talk about big data without mentioning the contributions from remote sensing technologies. As Academician Li Deren said, the diversity of imaging methods and the ability of remote sensing data acquisition has led to multi-source and massive remote sensing data [20]. This means that the time of remote sensing big data has arrived. The rapid development and extensive application of remote sensing technologies has gradually made remote sensing big data an important tool to study natural environment and social economy. Remote sensing big data also provide important support for the development of smart cities.

Remote sensing technologies include terrestrial remote sensing, aerial remote sensing, aerospace remote sensing and remote sensing from many other remote platforms (Table 4.2) [21]. Although the terminology remote sensing was invented in the 1960s, the development of the technology itself can be traced back to the 1820s. Particularly, the emergence of aerial remote sensing technology in the beginning of 20th century and the rise of aerospace remote sensing technology in 1960s have established a solid foundation for the current comprehensive remote sensing technologies. These technologies allow us to fully enjoy the advantages of remote sensing big data in smart cities in the new era.

**Table 4.2**   Categories of remote sensing technologies and typical models in China

| Primary | Secondary | Typical U.S. Systems | Typical Chinese Systems |
|---|---|---|---|
| Aerospace | Atmosphere RS | NOAA | FY-3 |
| Remote | Ocean RS | Sea Surveillance Satellite | HY-2 |
| Sensing | Earth RS | Landsat | ZY-3 |
| | High resolution RS | WorldView-3 | Gf-1 |
| | Other RS | SRTM | Remote Sensing Satellite |
| | Optical RS | UltraCam Eagle | space remote sensing |
| Aerial | Microwave RS | Airborne SAR | Airborne InSAR |
| Remote | Laser RS | Trimble-LiDAR | SW-LiDAR |
| Sensing | UAV RS | GoPro | DJI |
| Terrestrial | Close range photogrammetry | V-STARS | Lensphoto |
| Remote | Laser scan | Topcon GLS2000 | PROFILER 9012 |
| Sensing | Spectrum measurement | PSR+3500 | radiometer |
| | Ground monitoring | Dropcam HD | EVI-HD1 |

### 4.5.4.3 Remote sensing and smart cities

- Objectivity of remote sensing technology

Remote sensing technology is a kind of technology that acquires information and data by recording electromagnetic waves reflected or emitted by target objects [22]. Therefore, remote sensing data is a comprehensive representation of target objects' spatial forms, material compositions, physical characteristics and chemical properties. Because they are not subject to human impacts, remote sensing data have good objectivity. The objectivity of remote sensing data is crucial for applications.

For example, every day when we watch weather forecast on CCTV, we can see satellite images [21]. They contain remote sensing data recorded by China's Fengyun meteorological satellite and presenting national atmospheric conditions. Such data can give objective information on the atmospheric water vapour content, clouds thickness, motion trajectory and velocity of change, so that satellite meteorologists can analyse the probability of precipitation and its temporal and spatial distribution according to these data, and make weather forecasts and predictions on possible weather disasters. Another example is China's Number 3 resource satellite which can quickly deliver land remote sensing data for urban and rural constructions. This kind of data can objectively reflect urban and rural land using types, their spatial distribution and changes,

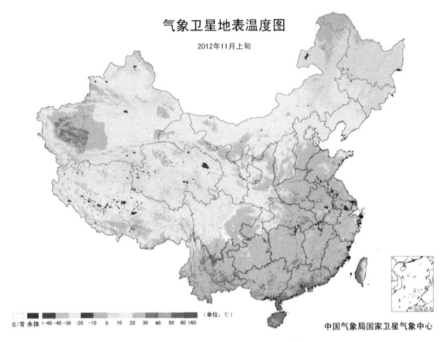

**Figure 4.13**    Weather forecast on CCTV.

and therefore can be used to quickly identify achievements and violations in urban and rural development.

- Multi-source of remote sensing technology

Remote sensing technology has many types, including a variety of remote sensing platforms, sensors, electromagnetic spectrums, imaging methods and of different resolutions. This can be characterized as the multi-source of remote sensing big data. The multi-source of remote sensing data can not only help us to deal with problems of natural and social environment from macro, meso and micro levels, but also to complement and to verify each other. Moreover, they can be integrated in applications in order to improve the speed and the quality of problem-solving. For example, the integrated space-air-ground observation network that China plans to build aims to enhance China's three-dimensional, dynamic and real-time remote sensing data acquisition capabilities, as well as the ability to provide accurate, integrated and continuous remote sensing data services for eco-environment protection and economic and social development. In addition, after the 2008 Wenchuan earthquake, scientific and technological specialists from China's remote sensing field took active actions to fully utilize China's aerospace, aerial and terrestrial multi-platform remote sensing capabilities. Obtaining multi-source remote sensing data has played an important role in rescuing, preventing further damage and assessing loss [23].

- Real-time of remote sensing technology

Despite aerospace, aerial and terrestrial, remote sensing data acquisition is the result of continuously recording information from target objects during the movement of remote sensing platforms or target objects. In other words, it is quasi-real-time information acquisition combined with real-time transmission. This allows us to get the latest data of target objects. The real-time feature of remote sensing big data is necessary for applications in various fields. Take forest fire as example, remote sensing monitoring can record real-time information on the timing, location, intensity and coverage of fire. Given that, rescuing and protection of forest resources will have reliable quantitative support. Moreover, we can also assess the damage caused by the fire. Real-time remote sensing monitoring of urban traffic flows enables us to acquire data on traffic and people flow. Analyzing these real-time monitoring data is helpful for real-time dispatch and optimization of road traffic and alleviating urban traffic congestion.

- High-resolution of remote sensing technology

With the development of remote sensing technology and the pushing of remote sensing application, the spatial resolution, temporal resolution and spectral resolution of remote sensing data have been increasing. In fact, high resolution is an important factor that brings about the time of remote sensing big data. Since the United States launched the first high-resolution commercial satellite Ikonos in 1999, countries that have strong space powers such as U.S. and China have been competing to launch satellites with increasing resolution and data quality progressively. For example, the QuickBird satellite launched by the U.S. in 2001 can achieve a spatial resolution of 0.61 meters. The GeoEye-1 satellite in 2008 has increased the spatial resolution to 0.41 meters. The WorldView-3 satellite in 2014 further improved the number to 0.31 meters. And finally, the GeoEye-2 satellite data projected to be launched in 2016-2017 will have a spatial resolution of 0.25 meters.

In China, the high-resolution earth observation satellite system currently under implementation will consist of high spatial, high temporal and high spectral resolution satellites.

| countries /institutions | name | Launch time | band | polarization | resolution m |
|---|---|---|---|---|---|
| USA | SIR-A/B | 1984 | L | HH | 7 |
| Canada | RADARSAT-1 RADARSAT-2 | 1995.11 2007.12 | C C | HH Quad | 10-100 2-100 |
| Germany | TerraSAR-X | 2007.6 | X | Quad | 1-16 |
| Japan | ALOSSAR | 2006.1 | L | Quad | 10 |
| ESA | ERS-1/2 | 1991-1995 | C | Quad | 10-1000 |
| China | GF3 | 2016 | C | Quad | 1-500 |

**Figure 4.14** Satellites with increasing resolution and data quality.

- Cyclic dynamics of remote sensing technology

Aerospace remote sensing platforms are based on various types of orbiting satellites. A very important feature of orbiting satellites is that they operate on previously set orbits constantly, and therefore the remote sensing data from orbiting satellites allow for displaying cyclic dynamics. For example, the U.S. land and resource satellite Landsat8 has a revisit cycle of 16 days. France SPOT6 of France has a revisit cycle of 26 days. China's satellite Resources One has a revisit cycle is 26 days and satellite Resources III 5 days. They all confirm the cyclic dynamics of remote sensing data. The cyclic dynamics of remote sensing data is very important for studying objects and phenomena that change dynamically, such as the growth of crops, the development of forest diseases and pests, and urban and rural construction. Meanwhile, the serial development of space remote sensing platforms has further lengthened the periodic monitoring and strengthened the dynamics of remote sensing data. For example, U.S. has launched a series of eight land and resource satellites since 1972 and has accumulated more than 40 years of dynamic data. China also has launched six Resource One series satellites since 1999, which helped to establish a solid foundation for the study of the long-term dynamics of resources and environment, as well as urban and rural construction. They also provide valuable information on studying China's urbanization and evolution, thus steadily supporting the future development of urbanization.

## 4.6 Conclusions

As mobile internet, cloud computing, big data, business models and other technologies and commercial applications continue to mature, the construction of smart city is also accelerating. The potential, prospect and impact of smart cities will become the new pursuit of China's urban construction.

This chapter has analysed and elaborated the construction and development trend of smart city construction. We have learned that due to the current urbanization, problems in our environment, resources and energy, and urban operations are slowly exposing. Traditional infrastructure and urban management cannot properly address these new problems in the development. Smart city thus provides an effective solution to these problems. The core of smart city construction is a series of new information and communication technologies, including Internet of Things, internet, and big data analysis. These technologies are applied in various aspects of urban life: government administration, transportation, health care, education, and environmental protection. We hope this chapter can shed some light on the future construction and development of smart cities, and provide certain reference and help.

## Appendix 1: Abbreviations

[C]
CCTV      China Central Television
CORS      Continuously Operating Reference System
CT        Computed Tomography
[E]
ESB       Enterprise Service Bus
[H]
HTML      Hypertext Mark-up Language
[I]
IoT       Internet of Things
IT        Information Technology
[L]
LTE       Long Term Evolution
[M]
MB        Mega Byte
[P]
PB        Peta Byte
[T]
TB        Tera Byte
[U]
UAV       Unmanned Aerial Vehicle
[X]
XML       eXtensible Markup Language

## References

[1] Nam, Taewoo, Theresa A. Pardo, "Conceptualizing smart city with dimensions of technology, people, and institutions," *Proceedings of the 12th annual international digital government research conference: digital government innovation in challenging times*, 2011, pp. 282–291.

[2] Yongmin Zhang, Zhongchao Du, "Present Status and Thinking of Construction of Smart City in China," *China Information Times*, vol. 2, 2011, pp. 28–32.

[3] Washburn, Doug, "Helping CIOs understand 'smart city' initiatives," *Growth*, vol. 17.2, 2009, pp. 1–17.

[4] Cheng Guo-Jian, "Industry 4.0 Development and Application of Intelligent Manufacturing," *Information System and Artificial Intelligence (ISAI), International Conference on. IEEE*, 2016, pp. 407–410.

[5] Jun Liang, HUANG Qian, "Opportunities and challenges in technology development from Digital City to Smart City," *Geomatics World*, Vol. 01, 2013, pp. 81–86.

[6] Palmisano, Samuel J, "A smarter planet: the next leadership agenda", IBM, 2008.

[7] Giffinger, Rudolf, "Smart cities Ranking of European medium-sized cities," 2007.

[8] FUJITA Masahisa, Krugman P R, Venables A J, *The spatial economy : cities, regions, and international trade*, MIT Press, 2001.

[9] RUAN Xiao-long and ZHAO Zhen-ying, "The Application of Cloud Computing on Intelligent City Construction," *Computer Knowledge and Technology*, vol. 32, 2014, pp. 7785–7793.

[10] Li Deren, "The Concept, Supporting Technologies and Applications of Smart City," *Journal of Engineering Studies*, vol. 04, 2012, pp. 313–323.

[11] Viktor Mayer-Schönberger, "Big Data: A Revolution That Will Transform How We Live, Work, and Think", Eamon Dolan/Houghton Mifflin Harcourt, 2013.

[12] Zou Guowei, "The application of big data technology to smart city," *Telecommunication Network Technology*, vol. 04, 2013, pp. 25–28.

[13] https://en.wikipedia.org/wiki/A/B_testing

[14] Ruan Xiao-long and Zhao Zhen-ying, "The Application of Cloud Computing on Intelligent City Construction," *Computer Knowledge and Technology*, vol. 32, 2014, pp. 7785–7793.

[15] LI Deren, "Application of High-resolution Earth Observation Technology in Smart City," *Journal of Geomatics*, vol. 06, 2013, pp. 1–5.

[16] Butler D, "2020 Computing: Everything, Everywhere," *Nature*, vol. 440, 2006, pp. 402–405.

[17] http://www.piesat.com.cn/alticle/PIE-GeoImage.html

[18] http://news.qq.com/a/20061012/001438.htm

[19] http://www.spacechina.com/n25/n144/n206/n214/c416213/content.html

[20] Chen K C, "Automatic Analysis and Mining of Remote Sensing Big Data," *Acta Geodaetica et Cartographica Sinica*, vol. 03, 2014, pp. 1211–1216.

[21] Dang An'rong, "Development of remote sensing data for smarter cities," *Construction Science and Technology*, vol. 03, 2016, pp. 15–18.

[22] Song Weijing, Liu Peng, Wang Lizhe, Lü Ke, "Intelligent Processing of Remote Sensing Big Data: Status and Challenges," *Journal of Engineering Studies*, vol. 3, 2014, pp. 259–265.

[23] http://www.nsmc.org.cn/NewSite/NSMC/Channels/FENGYUNImages More.html

## Biographies

**Wei-Dong Hu** received the Ph.D. degree in electromagnetic field and microwave technology from the Beijing Key Laboratory of Millimeter Wave and Terahertz Technology at the Beijing Institute of Technology, China, in 2004. He is currently an Associate Professor at the Beijing Key Laboratory of Millimeter Wave and Terahertz Technology, School of Information and Electronics, Beijing Institute of Technology. His research interests include electromagnetic, terahertz (THz) technology and terahertz remote sensing.

**Jianping An** received the B.S. degree from PLA Information Engineering University, Zhengzhou, China, in 1987, and the M.Sc. and Ph.D. degrees from the Beijing Institute of Technology, in 1992 and 1996, respectively, all in communication engineering. He joined the School of Information and Electronics, Beijing Institute of Technology in 1996, where he is currently a Full Professor. He is currently the Dean of the School of Information and Electronics with the Beijing Institute of Technology, Beijing. His research interests include the field of cognitive radio and signal processing in wireless communications.

**Shi Chen** received the B.S. degree in Physical Electronics from the Hefei University of Technology, Hefei, China, in 2015. He is currently pursuing the M.S. degree in electromagnetic field and microwave technology at the Beijing Institute of Technology. His research interests include terahertz remote sensing.

**Xin Lv** received the Ph.D. degree from the the Beijing Institute of Technology, China, in 1993, where he is currently a Full Professor. He is currently the chief of the Beijing Key Laboratory of Millimeter Wave and Terahertz Technology at the Beijing Institute of Technology, Beijing. His research interests include millimeter wave theory and technology, optoelectronic technology, antenna theory and design, electromagnetic scattering and stealth technology, millimeter wave system engineering and terahertz technology.

# 5

# Role and Importance of the Cyber Security for Developing Smart Cities in India

Vandana Rohokale[1] and Ramjee Prasad[2]

[1]Sinhgad Institute of Technology and Science, Pune, India
[2]Department of Business Development and Technology, Arhus University, Herning, Denmark

## Abstract

The first disruptive sociological wave came in between 1760s and 1830s. It was Industrial Revolution and it brought transition towards new manufacturing processes for factories, steam engines, electricity grids, etc. The second major sociological wave was the Political Revolution which brought the thought provoking political rebellion from Democracy versus Communism. The third and largest disruptive sociological wave struck around 1950s and 1970s was the Digital Revolution which changed the whole world with the massive usage of digital technologies like digital computing and information and communication technologies. For the progress of every nation, cities are playing an important role of a backbone for the economic growth. Around 54% of the world's population lives in urban areas and it is expected that it will reach around 66% by 2050. India's current population staying in urban areas is nearly 31% which contributes to 63% of India's gross domestic product (GDP) and it will reach around 50% by 2050 and will contribute 75% of India's GDP [1]. These numbers reveal the increased pressure on the current urban environment. This growth demands all-inclusive development of social, physical, institutional, and economic infrastructure. For improvement in the quality of life of the urban areas, smart city is the solution toward growth and sustainable development. But the ease of life kind of luxuries always comes with security and safety risks. Almost all cyber-physical systems are prone to cyber threats. A robust and secure smart city is the right of every citizen.

*Breakthroughs in Smart City Implementation,* 125–146.
© 2017 *River Publishers. All rights reserved.*

**Keywords:** Cyber Security, Smart Cities, Industrial Revolution, Political Revolution, Digital Revolution, Cyber Threats, Green Energy, Smart Village, Internet of Things (IoT), Smart Grid.

## 5.1 Smart City Introduction

For every country, cities are the growth engines of their future, offering their populations greater opportunities for education, employment, and prosperity. But, the negative effects of their growth can also result in traffic congestion, informal settlements, urban stretch, environmental pollution, exploitation of resources, and a major contribution to climate change. Efficient and intelligent technologies holds the answer to many of these urban challenges; i.e., "Smart Cities". The word Smart implies intelligence and in relation to Smart City it is nothing but the automation in the cyber physical systems (CPSs) of the city for bringing quality and ease of life to its citizens in the cost and resource efficient manner. Future population and environmental issues can be tackled with the sustainable development of the cities with long-term vision and mission. The main aim behind the development of cities is to provide resources to the citizens so that their productivity can be increased and they can live more suitable life. Because of continuous population growth, several problems are evolved for the smart city development in the developing countries like India. The key elements for the smart city development are the core infrastructure elements which include adequate and clean water supply, assured electricity supply provision, sanitation with consideration of solid waste management, health and education, sustainable green environment, efficient urban mobility and clean public transport, affordable housing for economically backward people, safety and security of citizens, and last but not the least, good governance with citizen participation. These core infrastructure key elements are depicted in Figure 5.1.

Some general attributes for sustainable smart city development are mentioned below.

1. **Area-based Development**: New Cities can be planned with a consideration of separate isolated green area and the residential projects can be implemented at a distant place from parks and green zone. Similarly, industrial zones should be isolated as well and more greenery should be incorporated in the industrial infrastructure plans. Already existing city infrastructures can be modified with the retrofitting techniques. Smart applications can be introduced in the existing cities.
2. **Housing and All-inclusiveness**: Affordable housing should be made available to economically backward community. Green and clean house

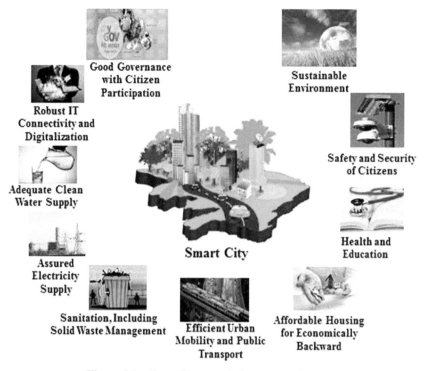

**Figure 5.1** Smart City: core infrastructure elements.

for all is the need of smart cities. Building rooftops can be mounted with solar cells to generate adequate amount of electricity required by individual home. Smart buildings with ultra-new smart and secure features from information and communication technology (ICT) are the need of time.

3. **Pedestrian and Bicycle Zones**: For congestion reduction, separate walkable localities can be developed and maintained with surrounding greenery. On the same lines, bicycle side roads which are distant from main roads can be strictly developed and maintained. This can help in reducing the $CO_2$ emission in the city and also help in building healthy and sustainable environment.

4. **Public Transport**: Clean and timely public transport system with electric vehicles, separate lines for trams, buses, and local trains are necessary. Usage of bio-fuels can help in reduction of $CO_2$ emission. Rapid transit systems will save lot of time and energy. Trains can be powered by electricity generated from renewable energy sources.

5. **Green Energy**: Smart Grid can bring automation and uniformity in the electricity generation, transmission, distribution, and usage. Mini grids can be implanted for individual buildings. Solar and wind farms can provide the major amount of electricity needed for individual city.

6. **Waste Management**: Recycle treatment plants for city's waste which can generate energy in turn is required. Centralized mechanical system can be developed for collection of dry and wet waste. Waste food can be utilized in the production of bio-gases.

7. **Health and Education**: Ecologically supported ecosystems reduce the pollution and health of the citizens can be maintained well. Also a centralized multi-speciality hospital facility with minor costs is the need of every smart city. For the comprehensive smart city development, citizen participation is very much essential. State of the art education techniques can develop smart people who can participate well in governance activities.

8. **Security and Safety**: Security of various CPSs and safety of the individual citizen especially elderly people and women is of prime importance. The information collected from various sources face lot of security issues such as rights, duties, and risks, etc. Cyber security mainly focuses on computing systems, data exchange media, and the actual information being processed.

9. **Worldwide Smart City Use Cases**: All over the world, the smart city movement is taking pace. Table 5.1 enlists some of the use cases of developed Smart Cities in the world with their attributes such as smart energy, smart building, smart mobility, smart people, and waste management [2].

**Table 5.1**   Some smart city use cases in the world

| S. No | Smart City | Territory | Smart Energy | Smart Building | Smart Mobility | Smart People | Waste Management |
|---|---|---|---|---|---|---|---|
| 1 | Masdar | Abu Dhabi And Arab Emirates | Photovoltaic and wind power plants | Green Buildings, Zero $CO_2$ emission, and renewable energy supply systems on building rooftops | Rapid transit system, shady streets for pedestrians, and Micro-metropolitan to semi-individual use | Research on technologies for efficiency, alternative energies | Recycle treatment plants for city's waste |

**Table 5.1** Continued

| 2 | Amsterdam | Netherlands, North Europe | Smart Grid Project, Mini-grids for individual buildings, green energy, flexible street lighting, and Ship to Grid project-73 electricity distribution units from renewable sources on river banks | Smart buildings with ceramic fuel cell plant for electricity provision, Smart school project for highest energy savings | Electric Vehicles, City Cargo for $CO_2$ emission reduction, clean public transport, cycle paths, canals | Citizens are flexible for new services and systems adaptation | Energy generation from recycle treatment of waste |
| --- | --- | --- | --- | --- | --- | --- | --- |
| 3 | Stockholm | Sweden | Green City, Bio-diversity and ecological support, Bio-fuel enabled city transport | Water purification, energy saving strategies and bio-gas for domestic applications | Walkable city, cycles, trams, trains, city buses powered from bio-fuels, trains powered by electricity generated from renewable energy sources, car use is reduced, and Ethanol is used as a fuel for vehicles | Noise reduction, Improvement in quality of life, 90% of the population lives less than 300 m from green area | Innovative waste treatment systems like Bio-gas plants from city's food waste, underground transport system for municipal city waste management |

(*Continued*)

**Table 5.1**    Continued

| S. No | Smart City | Territory | Smart Energy | Smart Building | Smart Mobility | Smart People | Waste Management |
|---|---|---|---|---|---|---|---|
| 4 | Freiburg | Switzerland | Smart energy sources, Energy production through fermentation of waste bio degradable products | Green Buildings | Bicycle, Tram or local city transport, Obtained European Local Public Transport Award | Freiburg Eco-station, courses and guided tours, Garbage theatre for children, Competitions and teaching units on "Ideas-No waste" | Energy production through fermentation of waste bio degradable products, full recycling of paper, plastic, organic materials, City uses 80% of recycled papers |
| 5 | Songdo | South Korea | Entire city fully wired with optical fiber, co-generation system for provision of natural gas and hot water to entire city | Largest private real estate investment, special concrete buildings to have great resistance to cold and heat | Songdo International Business District, Separate bicycle paths, pedestrian paths, Metro and bus lines, and charging stations for electric vehicles | Great awareness and participation in the business, and governance and all related activities | Centralized mechanical system for collection of dry and wet waste |

According to International Telecommunication Union (ITU), Smart City is defined as "A smart sustainable city (SSC) is an innovative city that uses information and communication technologies (ICTs) and other means to improve quality of life, efficiency of urban operation and services, and competitiveness, while ensuring that it meets the needs of present and future generations with respect to economic, social and environmental aspects."

European Union defines smart city as "A smart city uses information and communications technology (ICT) to enhance its liveability, workability and sustainability. In simplest terms, there are three parts to that job: collecting, communicating and "crunching." First, a smart city collects information about itself through sensors, other devices and existing systems. Next, it communicates that data using wired or wireless networks. Third, it "crunches" (analyzes) that data to understand what's happening now and what's likely to happen next."

The Government of India considers Smart city initiative as a Pan-city initiative wherein at least one smart solution is developed and it is replicated in the other cities. For the core infrastructure development, the government plans three models of area-based development. The first one is retrofitting wherein the existing physical systems will be retooled with digital infrastructure. The second model is redevelopment of the existing infrastructure by considering reconstruction. The third model in greenfield approach which considers the approach of building cities with completely new smart solutions.

## 5.2 Enablers of Smart City

Smart City is the integrated system of different CPSs and various innovative technologies such as ICT, IoT, ITS, Smart Grid, Cyber Security, etc.

### 5.2.1 Information and Communication Technology (ICT)

Information and Communication Technology acts as a prime enabler for making cities smart. Well-planning and execution of ICT infrastructure plays a vital role in the development of both Greenfield and Brownfield smart cities. ICT planning helps in using available resources in intelligent and efficient way resulting in cost and energy savings with reduced environmental footprints. For the successful and sustainable implantation of ICT for the development of smart city, the essential services include deployment of massive broadband networks, appropriate usage of smart devices and agents, developing urban smart spaces, developing web-based applications and e-services, and opening up Government data initiatives under right to information [3].

### 5.2.2 Internet-of-Things (IoT)

For transparent and seamless interconnection and integration of large number of heterogeneous CPSs and providing them various data subsets through different open data access services, IoT is the best technology. Urban IoTs support smart city development by using advanced communication technologies for supporting and providing value added services to the citizens so that they can actively participate and contribute to good governance. IoT can help in many tasks such as monitoring structural health of ICT infrastructure, city waste management, air quality monitoring, noise monitoring, traffic congestion, city energy consumption, smart parking, smart lighting, automation, and cleanliness of social spaces [4].

### 5.2.3 Smart Grid

Smart grid is the most effective long term approach in meeting the city's energy demands with cost efficient solutions. For sustainable smart cities, it is very much essential to provide assured electricity supply to every citizen and various ICT services with less environmental footprints. Use of renewable energy sources such as photovoltaic cells, wind power, tidal power, and vibration energy makes electricity grid smarter. Smart Water Meters can give information about usage, quality, leakage identifications, and preventive maintenance. For energy management, smart electricity meters will help in monitoring usage, energy efficiency improvement and reduction in losses and delays in fault detection and diagnosis in the distribution lines [5].

### 5.2.4 Intelligent Transportation System (ITS)

**Intelligent Transportation System** includes rapid transit system, shady streets for pedestrians, Micro-metropolitan to semi-individual use, Electric Vehicles, City Cargo for $CO_2$ emission reduction, clean public transport, cycle paths, canals. ITS demands walkable city planning, cycles, trams, trains, city buses powered from bio-fuels, trains powered by electricity generated from renewable energy sources, car use is reduced, and ethanol is used as a fuel for vehicles. ITS proposes to create corridors for automated vehicles to move city residents, electrify city fleets, and equip vehicles with connected vehicle technologies [6].

### 5.2.5 Cyber Security

"The Guardian" at the end of 2014 [7] in the interesting article reports that "The truth about smart cities: In the end, they will destroy democracy." The increased interfaces among various CPSs in the smart city paradigm globally interconnected physical, economic, social, and political sub-systems have increased the susceptibility of a smart city's cyber security. Due to the continuous big data transactions, cyber threats are getting multiplied. Security provision is a continuous process which gets modified according to previous attack experiences.

## 5.3 Smart Villages

Village is a heart of every developing country like India. Smart Village is a community-based movement conceived for providing benefits of ICT for the rural community. It is inspired mainly by philosophy and thoughts of Mahatma Gandhi with the aim of providing global means to the local needs. This initiative ensures access to sustainable energy, good education and healthcare, access to clean water, health, sanitation, and nutrition. The basic requirements of a village to be smart are depicted in Figure 5.2. Poverty and lack of basic needs is the main problem in front of most of the villages. The villagers need pucca houses to live in, hygienic food, water and clothes, awareness

**Figure 5.2**   Basic requirements of a Smart Village.

about money crops, various options for work and means of earning, clean, and strong metallic roads, usage of smart phones, and Wi-Fi and broadband Internet connectivity, smart cooperative service, etc.

People from the rural part of any country migrate toward urban areas because of the lucrative opportunities such as revenue generation, availability of good resources like work, education, health, lifestyle, etc. The cities can be decongested if all these facilities are provided in the village itself. Smart Village enables the growth of productive enterprise to boost per capita income with enhanced security and reliable social engagements [8]. Electricity is the basic need and main concern for development of smart villages. Many villages are installing their own solar power plants for small scale electricity generation. India is a fortunate country where throughout the year solar light and wind are available in ample amount. By making appropriate usage of these valuable non-conventional resources, electricity can be generated.

## 5.4 Cyber Threats to Smart City

Before discussing about security solutions, first we need to understand the risks. Let's consider a situation where with a small bug in the coding of technology-dependent services got shutdown. Citizens will feel frustrated if they experience non-functioning traffic control system, no street light and non-availability of public transportation. Imagine few hours without Internet connectivity can make the users feel helpless. Citizens will obviously not tolerate an adequate water or electricity supply, dark streets with no cameras [9].

Broadband Internet Connectivity and need of Wi-Fi-connected services expansion and use of autonomous sensors and associated logging and reporting devices are needed to remain always ON in the smart city scenario. This always on demand makes the small CPS prone to various cyber-attacks. Figure 5.3 shows the small CPSs per city which are needed to kept always ON for continuous monitoring and regulation of the services. These include energy distribution and management, street lighting, traffic control, public, and commercial transport, water distribution, waste management, parking regulation, and security, and surveillance systems [10].

Cyber-attack threat risk percentage of various CPSs in a Smart City is shown in Figure 5.4. Public Wi-Fi networks have the highest threat risk of attacks is almost 27% [11]. Smart Grid has the risk possibility of 18.6% and Transportation system has threat of almost 12.7% and video surveillance cameras pose 11.3% of risk [11]. This pie-chart is the eye-opener for the

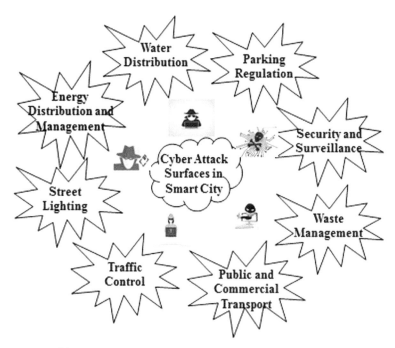

**Figure 5.3**  Possible Cyber-Attack surfaces in Smart City.

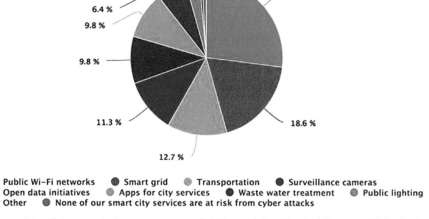

**Figure 5.4**  Cyber-attack threat percentage of different Cyber Physical Systems (CPSs) [11].

designers of smart cities. Robust cyber security solutions are a must for sustainable smart cities.

Various threats for the easily available data in Smart city include risk to big data, natural disasters situations, malicious activities, DDoS attacks can be purposely implemented on such systems. Also the privacy of the citizens is at risk due to open access to the data generated. Security-related issues to be given proper attention are privacy, network connectivity, complexity, security services, sensitive data organization, and availability of necessary data, emergency plan, and key management on large extent.

## 5.5 Security Challenges in Smart Cities

Internet-of-Things has become the heart of smart cities. Machine to machine (M2M) communication comes under the huge umbrella of IoT. Tiny sensors and RFIDs are the integral part of IoT and M2M. Both of these technologies are the basic building blocks of Smart Cities. These technologies are facing lot of security and privacy issues. Recently, number of IoT infrastructures has faced massive cyber-attacks. Smart refrigerators, television systems, smart grid sensors systems have been hacked. Many sensors in the IoT network have been compromised. IoT is struggling with many security and privacy challenges such as dynamicity and heterogeneity, security for integrated operational world with digital world, data security, device security, data source information, data confidentiality, trust negotiation, etc. [12].

For smart grid, three high-level security objectives are of prime importance including availability, integrity, and confidentiality. Timely and trustworthy access to smart grid infrastructure for getting information about electricity consumption or power grid is considered under availability. Lack of availability may lead in depreciation of the power delivery activities. Integrity is important for maintaining the authenticity and non-denial of the necessary information. Without integrity, it will become difficult for the central controlling unit to take correct power management decisions. For keeping intact the authorization of the private and proprietary information, confidentiality is a key factor. The major goal of confidentiality is not to leak the grid related secret to the public [13].

Safety, security, privacy, and trust are the major security challenges for providing cyber security in smart cities. IBM suggests IN3 model stating, smart city, its components, and its citizens are Instrumented, Interconnected, and Intelligent. Security and privacy issues rely on the usage of information within IN3. There are various sources through which information is generated in our

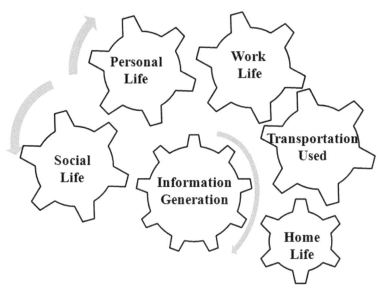

**Figure 5.5**   Information generation in everyday life of a citizen [13].

everyday life such as through social life, personal life, work life, transportation system used, etc. Large amount of data is generated through our everyday activities. Figure 5.5 mentions some of the attributes of information generation in every citizen's life. Citizen safety is very important aspect of cyber security of smart cities. Video surveillance is playing a vital role in our everyday life. But privacy, security, confidentiality, and availability should be maintained at the same time [14].

Among many of the critical challenges for smart city development are the few as mentioned below:

1. **Insecure Hardware**: Tiny Sensors used in IoT and almost all CPSs are prone to cyber-attacks because IoT devices are not standardized yet. Malicious users can compromise these sensors and may add malicious data causing system shutdown or signal failures, etc.
2. **Larger Attack Surface**: The data transaction among various CPSs in the smart city is not secure because it is either not at all encrypted or weakly encrypted. As shown in Figure 5.3, all these on systems like smart grid, ITS, street light, water distribution system, waste management systems, etc. need to be standardized with inclusion of appropriate level of encryption.

3. **Bandwidth Consumption**: In the smart city scenario, thousands of sensors or actuators will be surrounding us and trying to communicate either a server or another sensor. Normally, the sensors use unencrypted channel to communicate and there are many possibilities of hacking important information. These transactions create pressure on available spectrum by consuming considerable amount of it.

4. **Mobile APPs**: Nowadays, number of mobile apps has become an integral part of a human being's life. Localization, yoga, cooking, social networking, book reading, etc. apps have significantly changed every user's life. But some of these apps may be malware apps or app vulnerabilities may enter to your device through them [15].

## 5.6 India: Smart City Movement

On 28th January 2016, government of India announced the list of first 20 cities out of 98 shortlisted cities to be developed as smart cities under smart cities mission. The next 2 years ill see the inclusion of 40 and 38 cities, respectively. Out of these 98 cities, 24 are capital cities, another 24 are business and industrial centers, 18 are culture and tourism influenced cities, five are port cities, and three are education and health care hubs [16].

Indian Prime Minister Mr. Narendra Modi has a smart vision of developing hundred smart cities in India which will be capable of providing very high quality of life comparable with any European city. The Modi government revealed a plan to develop hundred smart cities across India and made an allocation of Rs. 100 Crore per city for next 5 years. Also, respective state governments will also provide Rs. 100 Crore for these cities for 5 years for the smart developments.

Metropolitan cities such as Delhi, Hyderabad, Surat, Coimbatore, Bangalore, Mangalore, Jamshedpur, Mumbai, and Chennai have launched project initiatives for deployment of advanced communication systems, metronetworks, intelligent transportation systems, smart meters, GPRS for solid waste management, online water quality monitoring, linking of Aadhar card to bank accounts, etc. Some of the smart city initiatives in India are summarized in Table 5.2. The construction for Delhi–Mumbai Industrial Corridor (DMIC) is already started which connects six states including Uttar Pradesh, Haryana, Rajasthan, Gujrat, Maharashtra, and Madhya Pradesh. It includes development of new seven cities on this way of the corridor. This project costs around US $90 Billion which has been already approved by Indian government [17].

**Table 5.2** Smart city initiatives in India

| Sr. No. | Smart City: Recent News Updates | Details of the Program Undertaken | Specific Tasks | Source |
|---|---|---|---|---|
| 1 | Smart Cities: Here's how Nagpur, Chandigarh, Pondicherry could evolve | With France proposing to develop Nagpur, Chandigarh, and Pondicherry as Smart Cities, they could evolve under a common urban model. | French companies like Alstom, Dassault, Egis, Lumiplan, RATP Transdev and Schneider have manifested interest in the Smart Cities Mission. Egis India, has already won a Smart City contract in Bhubaneswar and also bid in Chandigarh – separately it has earned an important contract for the Nagpur Metro. | [18] |
| 2 | SCC, NIUA and WD organize round table to discuss role of surveillance in India | The report is supported by Western Digital and talks about the need for surveillance to act as a catalyst for secure cities in India. It also talks about the role of data and IT in creating safer and smarter cities, as well as the urgent need for better storage capacities. | There is a positive correlation between urbanisation and crime. Therefore with affirmative evidence that surveillance does control crime. There is need for cameras that can assist in 'Crime detection' and 'Community watch'; while crime prevention, traffic management and women safety are other objectives that are deemed important. | [19] |
| 3 | Temples ring in high-end security systems (Sri Kashi Vishwanath did it!) | Talking about safety and security, [nowadays] even temples are not spared from the hit-list of terrorists. Hence the data availability for post event analysis is now paramount for all temple administration. | Varanasi-based Sri Kashi Vishwanath Mandir, which is always flooded with devotees, has installed CCTV surveillance cameras for crowd management, pilferage, and to prevent nuisance and damage done by the visiting crowd. | [20] |

*(Continued)*

**Table 5.2**   Continued

| Sr. No. | Smart City: Recent News Updates | Details of the Program Undertaken | Specific Tasks | Source |
|---|---|---|---|---|
| 4 | SCC roundtable on technology modernisation for safer and smarter city | Under the flagship "Safe City" project, the Union Ministry has proposed $333 million to make seven big cities—Delhi, Mumbai, Kolkata, Chennai, Ahmedabad, Bengaluru and Hyderabad | Since natural disaster disrupts development, there is a serious need of risk analysis by cities. An investment in risk sensitive development to prevent further loss of lives is essential. | [21] |
| 5 | Mumbai starts property mapping (first city to enable LiDAR project) | This is a first of its kind project which will create a unique platform for property tax administration, examining road surveys, mapping street furniture, hoardings and mobile telecom towers. | Use of Light Detection and Ranging (LiDAR) or laser for topographical mapping and contour generation is fast gaining acceptance worldwide. The commercial application of ground-based laser scanning has proven to be a fast and accurate technology for information extraction. | [22] |

## 5.7  Cyber Security Strategy for Smart Cities

Small bugs introduced in the coding related to the CPS's functionality may create huge impact like holding of passengers on board of flight or train by shutting down the track switching navigation system. The cyber security design systems should always stay ahead of cyber-attacks. Early warning detection of cyber breaches can lead to readiness for such attacks. Robust cyber security framework should be designed and developed by keeping eye on following things:

- Business Risk Understanding
- Secure Architecture Implementation
- Establishment of Incident response capabilities
- Awareness and skills improvement
- Third Party Risk Management

- Design and implementation of embedded and usable security
- Governance on the go implementation.

The cyber security strategy for implementation of smart cities should develop risk-based approach; clear priorities should be set, minimum requirement of the ICT baseline security should be properly defined. Threat and vulnerability information should be shared among all stakeholders. Incident response capabilities should be built with consistent and continuous approach. Public awareness should be developed through innovative education and work-force training on periodic basis. Public, private, academic, and government cooperation should be enabled.

## 5.7.1 India Cyber Security Policies for Smart Cities

Ministry of Urban Development of Government of India has developed a cyber-security model framework for smart cities and made it mandatory for the implementation of every layer of the generic architecture associated with smart city like sensing layer, communication layer, data layer, and application layer. This architecture needs to be open, interoperable, and scalable as suggested by National Institute of Standards and Technology (NIST) [23]. These are some of the main policies for provision of cyber security.

- Message exchange between various applications in the smart city must be fully encrypted and authenticated.
- Convergence of multiple heterogeneous CPSs should follow appropriate authentication and role based access control.
- City digital platform should be capable of communicating with various sensors and devices with heterogeneity.
- Information technology infrastructure in the smart city should follow the correct ISO standard.
- Safety and privacy of the confidential data should be maintained with proper encryption, authentication, and authorization techniques.
- Wireless broadband for the whole city should be provided with optical fiber cables.
- Sensors in the smart city should be robust and the traffic among various sensors should be encrypted.
- Data centers in the smart city should be equipped with advanced fire-walls, intrusion detection and prevention systems, denial of service prevention services, advanced persistent threat notification mechanism, well-equipped incident response teams, etc.

## 5.8 Summary and Outlook

Robust cyber security solutions are necessary to mitigate cyber-crimes and staying ahead of such kind of cyber-attacks. End to end encryption is necessary to be provided for CPS data transactions. Strong password policies can be deployed for various smart city services. Computing system firewalls and anti-virus systems are needed to keep updated. Auditing on periodic basis should be planned and executed with keeping records of audit logs. Trusted resources should be kept in isolation with public resources. Instead of considering security as a cost, the awareness should be developed in such a way that security should be considered as a plus for successful and secure smart city development.

## References

[1] Ministry of Urban Development, Government of India. (2015). *Smart City: Mission Transform-Nation*. "Mission Statement and Guidelines".

[2] Sanseverino, E. R., Sanseverino, R. R., Vaccaro, V., Macaione, I., and Anello, E. (2017). "Smart Cities: Case Studies", in *Smart Cities Atlas*, eds E. Riva Sanseverino et al. Berlin: Springer Tracts in Civil Engineering.

[3] Escher Group. (2015). *Five ICT Essentials for Smart Cities*.

[4] Zanella, A., Bui, N., Castellani, A., Vangelista, L., and Zorzi, M. (2014). *Internet of Things for Smart Cities. IEEE Internet Things J.* 1.

[5] Kimberly Klemm, K. (2015). *Smart Energy, Smart Grids and Smart Cities in Review*. Available at: http://www.energycentral.com/c/tr/smart-energy-smart-grids-and-smart-cities-review

[6] Kevin Dopart, K. (2016). *USDOT, Intelligent Transportation Systems Joint Program Office, "Beyond Traffic: The Smart City Challenge*. Available at: http://www.its.dot.gov/factsheets/smartcity.htm

[7] Ferraz, F. S., and Carlos André Guimarães Ferraz, C. A. (2014). "Smart City Security Issues: depicting information security issues in the role of a urban environment", in *7th IEEE/ACM International Conference on Utility and Cloud Computing*.

[8] Prabhu, M. J. (2016). Let's Create Smart Villages Before Building Smart Cities. *The Wire, Agriculture*.

[9] Bueti, C. (2015). *Overview of the activities of ITU-T Focus Group on Smart Sustainable Cities*.

[10] Cyberisk. (2016) "What Cyber Threats Are Smart Cities Facing?" Available at: http://www.cyberisk.biz/what-cyber-threats-are-smart-cities-facing/

[11] Help Net Security. (2016). *Smart cities face unique and escalating cyber threats*. Available at: https://www.helpnetsecurity.com/2016/10/20/smart-cities-cyber-threats/

[12] Rohokale, V. and Prasad, R. (2015). Cyber security for intelligent world with internet of things and machine to machine communication. *J. Cyber Security*, 4, 23–40.

[13] Rohokale, V. and Prasad, R. (2016). Cyber Security for SmartGrid – The Backbone of Social Economy. *J. Cyber Security* 5, 55–76.

[14] Elmaghraby, A. S. and Losavio, M. M. (2014). Cyber security challenges in smart cities: safety, security and privacy. *J. Adv. Res.* 5, 491–497.

[15] Assocham Security Report. (2016). *Cyber Security: A Necessary Pillar of Smart Cities*. India Security Conference.

[16] The Hindu. (2016). *Govt. Announces List of First 20 Smart Cities under Smart Cities Mission*. Daily News Internet Desk, Jan 28, 2016.

[17] Times of India. (2016). *Smart Cities Project is a mass movement:* PM Narendra Modi.

[18] SCC India Staff. (2017). *News*. Available at: http://india.smartcitiescouncil.com/category-news

[19] SCC India Staff. (2017). SCC, NIUA and WD organize round table to discuss role of surveillance in India. http://india.smartcitiescouncil.com/article/scc-niua-and-wd-organize-round-table-discuss-role-surveillance-india

[20] SCC India Staff. (2017). *Temples ring in high-end security systems (Sri Kashi Vishwanath did it!)*. http://india.smartcitiescouncil.com/article/temples-ring-high-end-security-systems-sri-kashi-vishwanath-did-it

[21] SCC India Staff. (2017). *SCC roundtable on technology modernisation for safer and smarter city*. Available at: http://india.smartcitiescouncil.com /article/scc-roundtable-technology-modernisation-safer-and-smarter-city

[22] SCC India Staff. (2017). *Mumbai starts property mapping (first city to enable LiDAR project)*. Available at:http://india.smartcitiescouncil.com/article/mumbai-starts-property-mapping-first-city-enable-lidar-project

[23] Government of India, Ministry of Urban Development. (2016). *Cyber Security Model Framework for Smart Cities*".

## Biographies

**Dr. Ramjee Prasad** is a Professor of Future Technologies for Business Ecosystem Innovation (FT4BI) in the Department of Business Development and Technology, Aarhus University, Denmark. He is the Founder President of the CTIF Global Capsule (CGC). He is also the Founder Chairman of the Global ICT Standardisation Forum for India, established in 2009. GISFI has the purpose of increasing of the collaboration between European, Indian, Japanese, North-American and other worldwide standardization activities in the area of Information and Communication Technology (ICT) and related application areas.

He has been honored by the University of Rome "Tor Vergata", Italy as a Distinguished Professor of the Department of Clinical Sciences and Translational Medicine on March 15, 2016. He is Honorary Professor of University of Cape Town, South Africa, and University of KwaZulu-Natal, South Africa.

He has received Ridderkorset af Dannebrogordenen (Knight of the Dannebrog) in 2010 from the Danish Queen for the internationalization of top-class telecommunication research and education.

He has received several international awards such as: IEEE Communications Society Wireless Communications Technical Committee Recognition Award in 2003 for making contribution in the field of "Personal, Wireless and Mobile Systems and Networks", Telenor's Research Award in 2005 for impressive merits, both academic and organizational within the field of wireless and personal communication, 2014 IEEE AESS Outstanding Organizational Leadership Award for: "Organizational Leadership in developing and globalizing the CTIF (Center for TeleInFrastruktur) Research Network", and so on.

He has been Project Coordinator of several EC projects namely, MAGNET, MAGNET Beyond, eWALL and so on.

He has published more than 30 books, 1000 plus journal and conference publications, more than 15 patents, over 100 Ph.D. Graduates and larger number of Masters (over 250). Several of his students are today worldwide telecommunication leaders themselves.

**Vandana Milind Rohokale** received her B.E. degree in Electronics Engineering in 1997 from Pune University, Maharashtra, India. She received her Masters degree in Electronics in 2007 from Shivaji University, Kolhapur, Maharashtra, India. She has received her Ph.D. degree in Wireless Communication in 2013 from CTIF, Aalborg University, Denmark. She is presently working as Dean R&D, in Sinhgad Institute of Technology and Science, Pune, Maharashtra, India. Her teaching experience is around 20 years. She has published one book of international publication. She has published around 30 papers in various international journals and conferences. Her research interests include Cooperative Wireless Communications, AdHoc and Cognitive Networks, Physical Layer Security, Digital Signal Processing, Information Theoretic security and its Applications, Cyber Security, etc.

# 6

# Smart City – Implementation Strategy

Arun Golas

Retd. Deputy Director General, Telecom Engineering Centre,
Department of Telecom, Government of India

## Abstract

Smart City is a buzz word currently, and all the countries are modernising their cities to implement newer services having fewer human interventions. Though everyone is still trying to figure out the best possible solution, it's not very clear as to what should be the strategy to implement various services for best results, more so because a large number of services are still evolving. Obviously, the solution, as usual, shall not be 'one size fits all', and the development pace would also be different in different cities depending upon various degrees of their readiness. This article highlights various probable issues to be kept in mind while implementing the project.

**Keywords:** Smart City, Improving quality of life, Optimal utilisation of resources, Use of green technologies, Dynamic management of parameters and resources, Billions of interconnected devices, Embedded sensors and actuators, Ubiquitous network of networks, Distributed network control, Heterogeneous data, User-controlled parameters, Convergence of wireless and wire-line, Optical Fibre Cable-based transport, Shift of focus to 'over-the-top (OTT) service provider', Smart City Implementation strategy, Coordinated development of cities, Transparent and citizen-friendly governance, Active involvement of all citizens, Disciplined behaviour, Smart City Framework – Technology, Human, and Institutional Frameworks, Smart City Platform and Technology, Smart City Road-map, Internet of Things, Data management, Data Security and Privacy.

*Breakthroughs in Smart City Implementation,* 147–164.

## 6.1  Glimpses into Future Life

All of us dream of comfortable life in future, and some ideas could well be pure fantasies at this stage. But, we should not forget, that flying in air was a fantasy a century away, or carrying a computer in our pocket was a dream few decades ago. Let us have a look at some of the future scenarios.

a. On a typical morning, the curtains open to let sun in, and the radio switches on for music to wake you up. As you get-up and while you freshen-up, news from your favourite channel is projected on wall to update you. Subsequently, music is played as you commence your daily exercise. Obviously, your health parameters are continuously tracked to make any required modification in the exercise regimen, and you get a feedback as soon as you complete your scheduled daily quota of the exercise, or if you fail to do so. After the exercise, as you enter the bathroom, the lights start flashing with music. As you turn the shower faucet, the water temperature is adjusted to your particular preference. When you open the wardrobe, a specific dress may be suggested for the special meeting scheduled for today. When you take your favourite cereal out of the fridge, replenishments are also ordered automatically in the background. You may either pick up the supplies from the store, or else they can be delivered to you, as per your preference. As you pick up your car keys, your friend can be intimated, if you want so.

b. During a typical commute to the office, the navigation screen tells you to pick up the fastest, not necessarily the shortest, route. You are also intimated of the traffic conditions dynamically, to enable you to change over to an alternate route, in case of any hold-up on the initially selected route. With the autonomous cars, this activity will be taken over by the vehicle itself, and you can either relax in the seat, or prepare for the day. The office door unlocks after authenticating your attributes, and it opens to let you in. All your other attributes are also registered in the system for further analysis. The screen at your desk brings up the schedule for the day with priorities, if any.

c. A typical shopping experience will also change as there would be no need to visit shops or showrooms to try-out clothes, and you would be able to see, right in the comfort of your house, as to how the dress would look on you. If you don't like the fabric colour or design, you would be able to see other options. Even if there are fluctuations in your weight, you would not have to try different sizes, as your size parameters would be updated dynamically. Before we finalise, most of the times, we also want to have a feel of the texture of fabric. With introduction of tactile controls, this would be also possible, without going to the store. The finalised piece will be delivered to your door-step.

All these typical Sci-Fi scenarios are just a few glimpses into the future, and converting them into reality is not too distant in future. In fact, some of the services are already in use in not only the cities of developed countries, but also of developing countries. Obviously, with such facilities, our daily life would be simplified, and we would be in a position to work more efficiently, thereby improving the quality of our lives. However, whatever we are supposed to do ourselves would still have to be done by us. The machines would be designed to assist us and not to replace us.

It would be relevant to keep in mind that a large amount of parameters about every person would be gathered and recorded for different applications. For example, different sets of health parameters collected would be used differently by different agencies. A health facility may monitor the vital parameters to keep track of the health condition, and if the need arises to intimate the person for any preventive action, to avoid any emergency, or before the condition worsens. Moreover, the periodicity of monitoring would vary according to the general health of an individual. For elderly persons or unwell persons periodicity of analysing the data would be much higher. A house itself may monitor another set of parameters to maintain temperature and humidity, as well as suggest and order nutritious food and supplies, etc. Another set of parameters would be required for keeping track of the clothing and shoe sizes, etc. The employer may also keep a track of health parameters, as well as the exercise routine adhered to, to analyse if any corrective action is required to avoid absenteeism, and to design suitable reward systems.

As power would be required by the all such connected devices, the overall requirement of power would go up exponentially. Hence, an important inescapable aspect of the network architecture would have to be to minimise the power being consumed by various devices. Unless required to transmit or receive signal, the devices shall have to remain dormant consuming a very low level of power, and preferably, no power at all. Similarly, all natural resources shall have to be used optimally. Instead of brick-mortar outer walls, the option would be to have outer walls made up of only glass, to let the natural light in, eliminating the use of light-bulbs during the day. Alternatively, the house architecture might also revert to the older styles with an atrium in the middle, to let natural light come into various rooms.

## 6.2 Smart City

In any Smart City, all its constituents will be smart, which implies that all the services and various infrastructures will be monitored (Figure 6.1). Considering dynamic changes in almost all conditions, it will be essential to perform live monitoring of various parameters around us in order to optimise the utilisation of various resources and services, such as, household management, energy supply, water supply, waste management, schools, libraries, hospitals, plants and trees, transportation systems, governance, law enforcement, and various other community services. As optimisation of resources is going to save a massive amount of money, every administration would like to have only smart cities.

However, the most important question for the administrations is whether is it possible to 'build' a Smart City right away? And the plane answer is of course not. This goal cannot be attained without creating Smart Homes, and before that having Smart Persons! The most difficult facet to make us smart is to bring about definite behavioural changes. First of all there has to be acquisition of soft-skills to use the facilities, and a free will to follow rules. Let us admit the fact that we are highly unwilling to follow any discipline, and the entire concept of smart person, smart homes, and smart city is based

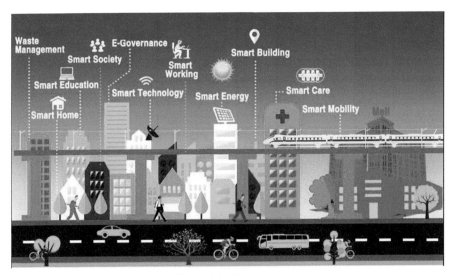

**Figure 6.1**   Smart city.

*Source*: Architect corner: Top 10 smart city start-ups in 2017.

on almost non-flexible disciplines. With no exceptions, all the citizens shall have to follow the laid down guidelines, rules, and regulations.

To make all these services feasible, various devices, objects, packaging, and gadgets would have embedded sensors, and a few of them would have actuators as well. All such sensors and actuators would be interconnected through suitable networks.

The basic challenge is to interconnect all the devices through a reliable network. Not only the number of devices is mind-boggling, but also the ubiquitous network shall have to be made available at all times. It is expected that 50 billion devices will have to be interconnected by 2050! However, some people believe that it will not take as long, and 50 billion devices will have to be connected as early as 2020.

## 6.3 Network Configuration

As there would be different types of devices interconnected at peer and higher levels, there would be a complex network of networks. Each network would be performing a specific task of collecting and processing data. As many of the actions to be performed will be localised, it may not be necessary to carry the data to the central control for decision-making, hence distributed intelligence will have to be built in the system.

The data generated will be heterogeneous and hence, an all-packet omni-present massive global network is the only solution. However, end-to-end Quality of Service is imperative as any degradation in quality may lead to catastrophe. Simultaneously, the network parameters shall have to be user controlled, to ensure that the data channels are used optimally all the times.

With billions of connected devices, the aggregate data may shoot up to petabytes, implying the requirement of massive networks. However, all the data from all the sensors does not have to be carried up to a central server. Majority of the data will be processed right in the home or a nearby facility. Moreover, despite the huge number of sensors, all of them will not be transmitting all the time, nor would they be sending gigabytes of data. Hence, the required data carrying capacity would be much lesser than the apparently needed one. Notwithstanding such scenarios, as the data requirements are spiralling up, despite innovations and improvements, radio networks will still be unable to meet the data deluge, and it will be essential to tap the wire-line systems, comprising copper and optical fibre. With the advancements in optics, each fibre-pair is capable of carrying bi-directional data speeds of over 70 Terabit per second [1]. Such speeds are unheard of in radio communication, and even if possible in future, would still be left behind with corresponding advancements in optical communication. As each optical fibre cable has multiple fibres, the capacity increases multi-fold, akin to having the impossible advantage of multiple spectrums of same frequencies! Hence, for optimisation of networking resources, convergence of wired and wireless is the only option.

With such high amount of data generated at the user's end, the shift of focus from telecom service provider towards the over-the-top (OTT) service provider is already evident. In the years to come, there would be a paradigm shift as the focus would completely be on the service provider only, the telecom services moving completely to the background, and the present telecom service providers would practically become the network providers, albeit a few who would also enter the arena of provisioning of new services.

## 6.4 Implementation

Though the opinions on the definition of Smart City are divided, but in the current context, a quite suitable definition seems to be "A smart city is an urban development vision to integrate multiple Information and Communication Technology (ICT) and Internet of Things (IoT) solutions in a secure fashion to manage a city's assets, which include, but are not limited to, local departments' information systems, schools, libraries, transportation systems, hospitals, power plants, water supply networks, waste management, law enforcement, and other community services".

"The goal of building a smart city is to improve quality of life by using urban informatics and technology to improve the efficiency of services and meet residents' needs. ICT allows city officials to interact directly with the community and the city infrastructure and to monitor what is happening in the city, how the city is evolving, and how to enable a better quality of life. Through the use of sensors integrated with real-time monitoring systems, data are collected from citizens and devices – then processed and analysed by the IoT networks. The information and knowledge gathered are keys to tackling inefficiency" [2].

World-over, the administrations are taking steps to make their cities smart. Indian administration has also decided to convert, in the first phase, its 100 cities into smart cities through city improvement (retrofitting), city renewal (redevelopment) and city extension (greenfield development), plus a pan-city initiative in which Smart Solutions are applied covering larger parts of the city [3]. All-round development is imperative as the benefit of smart solutions shall be available only then, and integration with the rest of the world shall be possible only then.

a. Make governance citizen-friendly and cost-effective – increasingly rely on online services to bring about accountability and transparency, to reduce cost of services and providing services without having to go to respective offices. Apply Smart Solutions to infrastructure and services in area-based development to make them better. Create and refurbish road network for vehicles, cyclists, and pedestrians. Promote a variety of public transport options to provide long haul and last mile connectivity. Reduce traffic congestion, air pollution, resource depletion, and boost local economy, promote interactions and ensure security.

b. One of the biggest challenges in implementation of smart solutions is the cost involved. No expert analysis is required to deduce that for every quantum of solutions added, there would be proportionate addition to the costs. Hence, the mix of solutions shall have to be judiciously decided on case-by-case basis, depending on the starting level of development of the city, with no single strategy fitting all.

c. Judicious use of green technologies has to be an integral part of the project. The design of the service has to be around green technologies, rather than finding ways in making it green. To counteract the harmful effects of the climate-change, it is essential to avoid all wastages of energy, fuel, food, water, etc. Even though increasing number of newer technologies shall be used in future, they will work in harmony in such a way that the overall energy usage shall be reduced. In the same spirit, it must be ensured that no vegetation is unnecessarily destroyed, and if destroyed, must be replenished suitably. Smart Cities shall have to coexist with nature, and aid in restoration of ecological balance.

d. Presently, there is tremendous amount of chaos and disorder in normal lives in India. Accordingly, there are a large number of behavioural issues which shall hinder the steps taken for the changes. Suitable solutions shall have to be worked out to overcome them to bring order. Simultaneously, the citizens have to be convinced that the changes are for their betterment and can be brought about only with their active participation.

e. As the services shall become more and more network-based, permanent solutions for the major problem of frequent power outages shall have to be worked out.

General perception of use of civic facilities through brute-force shall have to change; provision of services would be possible only if pre-defined discipline is followed, as all the reactive measures would be rule-based with almost no exceptions. Accordingly

a. Suitable deterrents would be in place to avoid any encroachment on government property. Failure to do so may result in removal of possession and heavy penalties. Similarly, wilful damage to public or private properties shall entail severe penalties. All unruly behaviour shall be liable to be penalised.
b. All vehicles shall have to be parked in the areas earmarked for the purpose. Failure to do so would entail imposition of fine which shall automatically be deducted from the defaulter's account.
c. Similarly, failure to observe all other rules and regulations, littering, etc., would bear similar consequences.

## 6.5 Framework

Any Smart city will have a mix of ICT and IoT technologies for the provision of various services. Accordingly, the network architecture would be technology centric. However, implementation strategy and framework could be notionally different depending on the mix of services provided and the mix of focus attribute of the city. Whatever framework is selected, it has to be sustainable, leading to harmonious growth of the city.

### 6.5.1 Technology Framework

#### 6.5.1.1 Digital city

A Digital City comprises a connected community that combines broadband communication infrastructure; a flexible, service-oriented computing infrastructure based on open industry standards; and innovative services to meet the needs of government and their employees, citizens and businesses [4].

The main purpose is to create an environment in which citizens are interconnected and easily share information anywhere in the city.

#### 6.5.1.2 Virtual city

In a Virtual City, functions are implemented in a cyberspace; it includes the notion of hybrid city, which consists of a reality with real citizens and entities, co-existing with a parallel virtual city of real entities and people. In practice this idea is held up through physical IT infrastructure of cables, data centres, and exchanges.

### 6.5.1.3 Information city

Information City collects local information and delivers them to the public portal. The inhabitants would use Internet for their daily life and work as well, as they can obtain all information through ICT infrastructure. Economically and socially speaking, it can be an urban centre with linkages among civic services, government institutions, and people [5].

### 6.5.1.4 Intelligent city

Intelligent City involves research or technological innovation to support learning and innovation procedures. Knowledge, learning process, and creativity shall have great importance and the human capital would be the most precious resource. Every infrastructure would also be updated continuously [6].

### 6.5.1.5 Ubiquitous city (U-city)

Ubiquitous City creates an environment that connects citizens to any services through any device [7]. It would be an extension of digital city providing accessibility to every infrastructure through any device at any time. U-city would have processors embedded in urban elements, whereas the virtual city creates virtual space of the real elements.

### 6.5.2 Human Framework

### 6.5.2.1 Creative city

Creativity is recognized as a key driver in a Creative City. Social infrastructures, such as intellectual and social capital, would be indispensable factors. Benefits would be derived from people and their relationships, mix of education and training, culture and arts, business and commerce [8].

### 6.5.2.2 Learning city

Learning city is involved in building skilled workforce [9]. Being individually proactive, it would improve the competitiveness in the global knowledge economy, aiding in formation of city cluster, one-to-one link between cities, etc.

### 6.5.2.3 Humane city

Humane City exploits human potential, the knowledge workforce, focussing on education to build a centre of higher education, to further increase the level of knowledge [10]. As such places would act as a magnet for creative people and workers, smart cities would get smarter; while other places would be left behind.

### 6.5.2.4 Knowledge city

Knowledge City is related to knowledge economy and innovation process, similar to other cities, except that it relies more on knowledge economy and innovation to boost growth [11].

### 6.5.3 Institutional Framework

Members of Smart Communities share their interest, and work in a partnership with government and other institutional organisations to push the use of ICT to improve quality of daily life. A smart community would make a conscious and agreed-upon decision to deploy technology as a catalyst in solving its social and business needs which otherwise would be quite demanding [12].

## 6.6 Platforms & Technologies

Information and Communication Technologies, Cloud-based services, Internet of Things, semantic web, real-world user interfaces, and smart phones, smart meters, networks of sensors and RFIDs, and actuators, open new ways to collective action and collaborative problem solving.

Online collaborative sensor data management platforms will allow sensor owners to register and connect their devices to feed data into an on-line database for storage and allow developers to connect to the database and build their own applications [13].

Connecting buildings, infrastructure, transport, networks and utilities, will offer a physical space for experimentation and validation of the IoT functions.

Electronic cards, possessing unique encrypted identifier, will allow owners to log in services without setting up multiple accounts.

## 6.7 Internet of Things

Definitely, implementation of Smart Cities would be impossible without Internet of Things (IoT). It is the inter-networking (Figure 6.2) of physical devices, connected devices, smart devices, buildings, network connectivity, and other items – embedded with electronics, software, sensors, actuators, and that enable these objects to collect and exchange data.

**Figure 6.2** Internet of Things.

*Source*: Computer business review.

A global infrastructure for the information society, enabling advanced services by interconnecting (physical and virtual) things based on existing and evolving interoperable information and communication technologies. Through the exploitation of identification, data capture, processing and communication capabilities, IoT makes full use of things to offer services to all kinds of applications, whilst ensuring that security and privacy requirements are fulfilled. From a broader perspective, the IoT can be perceived as a vision with technological and societal implications.

IoT is expected to greatly integrate leading technologies, such as technologies related to advanced machine-to-machine communication, autonomic networking, data mining and decision-making, security and privacy protection and cloud computing, with technologies for advanced sensing and actuation [14].

### 6.7.1 Network Configuration

As all items or packaging and gadgets would be connected, they all will have sensors and actuators embedded in them, and these would be interconnected through suitable networks all-packet network omni-present massive global network. Thus, there would be integration of real world with interconnected computer based systems.

Hence, a major design challenge is to build a modular architecture that can keep on growing and evolving to provide seamless interconnectivity, as newer heterogeneous objects and services – hereto unknown, unheard, and unimaginable – will be added to the eco-system, all the time. The architecture should also have such resilience so as to keep on of functioning even after disengaging older sensors, nodes and applications, without any disruption in services. Unless the laid down standards for every service is followed, such scenarios would be impossibilities to achieve. This further necessitates distributed intelligence at the edge-devices architecture, as well as to avoid flow of ever-increasing amount of data on the internet backbone [15].

With the interconnection of billions of sensors and actuators, a humongous amount of data would be generated. If the data is stored in cloud for further processing, not only massive storage capacities would be needed, but also there would be a tremendous load on the transmission network, necessitating bandwidths of petabits per second. This architecture would be impractical and unfeasible. Hence, instead of carrying all the data to a centralised application on internet backbone, it would be essential to resort to fog computing with distributed intelligence. With this, it would be possible to take quick decisions and actions for the simple and routine issues, locally – preferably within the house, or the logical extreme. After taking action at local end, if further action is also required at some other end, only the event based relevant data, will be transferred upward; rather than sending the entire raw data. The necessity to carry huge amount of data to the cloud would no longer be necessary, cutting down the bandwidth requirement drastically. However, network architecture must have to provide parameters on case-by-case basis controlled by the user – human and non-human – to ensure end-to-end Quality of Service.

Considering the fact that a large number of applications shall require data for the same parameter, the data structure shall have to be harmonised by collaborative efforts by the application developers, to enable seamless data sharing.

Harmonised man-machine interfaces too shall have to be developed to avoid the problem of having "basket of remotes", when there are different remotes for different applications [16]. One of the solutions could be through "predictive interaction", where cloud or fog based decision-makers will predict user's next action, and trigger an appropriate reaction [17].

Another big challenge to be met with the generation, transfer and processing of the huge amount of data, is its security and privacy [18].

a. Data can be hacked at any of stages, when it is being collected, transmitted, stored, or processed. The network architecture has to be built around security, rather than devising solutions for data security, to thwart all attempts of eavesdropping, stealing, or mutilation of data during transmission, in storage, or when being processed. At the time of designing, it should be ensured that data collection, storage and processing would be secure at all times. A "defence in depth" approach should be adopted, and data should be encrypted at each stage [19].

b. As a large number of transmitters are sending out a large amount of data aimed at different collecting devices, it can also be picked up by a large number of unauthorised devices. Hence, it would be essential that the data is not collected unless there is an explicit consent given by the user to that device or application. Presently, the 'consent' for data sharing is a mere formality, and it is the 'demanded' prerequisite for the usage of most of the applications.

c. For an application to give desired results, a definite set of parameters is required to be collected. However, at present, data pertaining to other unrelated sets of parameters are also collected. It is essential that only the minimal agreed set of data is collected, to honour the privacy of the user, notwithstanding the fact that presently the users merrily share almost all their personal data online to various applications, without realising that they themselves are unwittingly, yet willingly, undermining their privacy. Notwithstanding any such situation, only minimal data shall be collected and stored for a limited time only; and the amount of data no longer required, shall be deleted from all locations.

### 6.7.2 Devices

In addition to the presently used devices, such as phone, personal digital assistant, tablet, computer, television, almost everything else around us would be connected through IoT!

## 6.8 Roadmap

The first step, for implementation of Smart Solutions, is to define exactly various characteristics of the community related to geography, links between cities and countryside, and flows of people between them. It is essential to analyse citizens and communities, know the processes, business drivers, create policies, and objectives to meet the citizens' needs.

It is essential to determine the benefits of converting the city to a smart one, study the community to know the citizens, the business's needs – know the citizens and the community's unique attributes, such as, age of the citizens, their education, hobbies, and attractions of the city.

A definite 'Smart City Policy' has to be developed to drive the initiatives, define roles, responsibilities, objectives, and goals. Based on this policy, plans and strategies shall to be created as to how the goals will be achieved.

It is essential to engage citizens through the use of e-government initiatives, fairs and exhibitions, open data, sport events, etc. Active involvement of all the citizens shall be the key to the success of any initiative. Different degree of efforts would have to be made to educate the citizens about the services and in imparting training to use them.

People, Processes, and Technology (PPT) are three principles of the success of a smart city initiative. This requires a holistic customized approach that accounts for city cultures, long-term city planning, and local regulations. Technology can be implemented to meet the citizens' needs to improve quality of life, and create real economic opportunities [20].

## 6.9 Conclusion

Earlier, it was a miracle to talk to someone living far away, and in a short time it has become possible to communicate with any person, anytime, anywhere! Shortly, it would be possible to communicate with him on any screen. And in future anything would be communicating with anything, anytime, anywhere, on any screen!

India is a diverse country; different cities have different ethos, which need different strategies for implementation of same service. Different sets of services will have to be to be implemented in different cities. This necessitates different city-specific trajectories across the country.

Whatever framework is chosen, and whatever platform and technologies are shortlisted, the objectives would be met only when a holistic view of the entire city is kept in mind. It implies that all the activities shall have to be in the direction of making the citizens smarter before making the city smarter, and then only, the citizens of the entire city would be in a position to reap its benefits, in a continuous and sustainable manner.

Finally, it is extremely important to keep in mind that the Smart Cities are meant to enrich the lives of human beings by improving the quality of their live, and not to replace them by machines. Hence, the human beings shall have to continue to do whatever they are supposed to do themselves.

# References

[1] Aron, J. (2013). Big data, now at the speed of light: Information superhighway approaches light speed. *New Scientist* 14.

[2] Musa, S. (2016). *Smart City Roadmap*. Available at: http://www.academia.edu/21181336/Smart_City_Roadmap

[3] Government of India *Smart City Mission*. Available at: http://smartcities.gov.in/content/innerpage/what-is-smart-city.php

[4] Yovanof, G. S., and Hazapis, G. N. (2009). *An Architectural Framework and Enabling Wireless Technologies for Digital Cities & Intelligent Urban Environments*. Available at: https://link.springer.com/article/10.1007/s11277-009-9693-4

[5] Wikipedia *Smart City*. Available at: https://en.wikipedia.org/wiki/Smart_city

[6] Komninos, N., and Sefertzi, E. (2009). *Intelligent Cities*. Available at: http://www.urenio.org/wp-content/uploads/2008/11/Intelligent-Cities-Shenzhen-2009-Komninos-Sefertzi.pdf

[7] Anthopoulos, L., Fitsilis, P. (2009). "From online to ubiquitous cities: the technical transformation of virtual communities," in *Next Generation Society. Technological and Legal Issues: Lecture Notes of the Institute for Computer Sciences, Social Informatics and Telecommunications Engineering*, Vol. 26, eds A. B. Sideridis and C. Z. Patrikakis (Berlin: Springer). Available at: https://link.springer.com/chapter/10.1007/978-3-642-11631-5_33

[8] Bartlett, L. Available at: engagingcommunities2005.org

[9] Moser, M. A. (2001). *What is Smart about the Smart Communities Movement?* Available at: http://www.ucalgary.ca/ejournal/archive/v10-11/v10-11n1Moser-browse.html

[10] Glaeser, E. L., and Berry, C. R. (2006). *Why Are Smart Places Getting Smarter?* Available at: http://ksghauser.harvard.edu/index.php/content/download/70209/1253646/version/1/file/brief_divergence.pdf

[11] Dirks, S., Gurdgiev, C., and Keeling, M. (2010). *Smarter cities for smarter growth*. Available at: https://www.zurich.ibm.com/pdf/isl/infoportal/ IBV_SC3_report_GBE03348USEN.pdf

[12] Eger, J. M. (2009). *Smart growth, smart cities, and the crisis at the pump a worldwide phenomenon*. Available at: http://dl.acm.org/citation.cfm?id=1552016

[13] Boyle, D., Yates, D., and Yeatman, E. (2013). *Urban Sensor Data Streams: London 2013*. Available at: http://ieeexplore.ieee.org/document/6576751/

[14] International Telecommunication Union (2012). *ITU-T Y.2060 "Series Y: Global Information Infrastructure, Internet Protocol Aspects and Next-Generation Networks"*. Geneva: International Telecommunication Union.

[15] Garcia Lopez, P., Montresor, A., Epema, D., Datta, A., Higashino, T., Iamnitchi, A., et al. (2015). *"Edge-centric Computing: Vision and Challenges,"* in *Proceedings of the Newsletter ACM SIGCOMM Computer Communication Review*, (New York, NY: ACM).

[16] Gassée, J.-L. (2014). *Monday Note, "Internet of Things: The 'Basket of Remotes' Problem"*. Available at: https://mondaynote.com/internet-of-things-the-basket-of-remotes-problem-f80922a91a0f

[17] Intel (2015). *Better Business Decisions with Advanced Predictive Analytics*. Available at: Intel's site.

[18] Mason H., and Curran (2016). *The 'Internet of Things': Legal Challenges in an Ultra-connected World*. Available at: https://www.mhc.ie/latest/blog/the-internet-of-things-legal-challenges-in-an-ultra-connected-world

[19] Brown, I. (2015). *Regulation and the Internet of Things*. Oxford: Oxford Internet Institute.

[20] Wikipedia *Smart City*. Available at: https://en.wikipedia.org/wiki/Smart_city

## Biography

**Arun Golas** worked in Department of Telecom (DoT) from 1981 to 2013. He retired from the post of Deputy Director General (DDG) in Telecom Engineering Centre (TEC), New Delhi, in the grade equivalent to the Additional Secretary to the Government of India.

He holds a B. Tech. degree in Electronics Engineering from IIT Kanpur.

He played key role in formulation of various telecom policies by DoT during 2007–13, for standardisation, certification, promotion of indigenous research and development, manufacturing and deployment of telecom products.

He worked in TEC for 11 years, handling the standardisation activities for satellite communication, optical communication equipment.

He played a significant role in the WTDC-2010, Hyderabad by drafting 'Hyderabad Declaration', and chairing its drafting committee. 3 proposals contributed by him were included in the ITU-D resolutions. He has also contributed in APT, ITU-T, TSAG, GSC, etc.

He has a rich experience of over 32 years in various fields of telecom, viz., standardisation, testing and certification; installation, operations and maintenance of telephone exchanges; planning and development of switching, transmission, and computer networks; planning and deployment of districts level mobile communication network; designing and construction of telecom buildings; managing staff; development of training courses, training of manpower, training of trainers; material management; etc.

# 7

# Challenges in the Design of Smart Vehicular Cyber Physical Systems with Human in the Loop

Agata Manolova, Vladimir Poulkov and Krasimir Tonchev

Faculty of Telecommunications, Technical University of Sofia,
8 blv. Kliment Ohridski, Sofia 1756, Bulgaria

## Abstract

Cyber-Physical Systems (CPSs) are mainly related with the control and monitoring of physical environments and phenomena through sensing and actuation systems consisting of distributed computing and communicating devices. CPSs have seen a great expansion with enormous societal and economic impact facilitating various services such as medical systems, assisted living, traffic control and safety, advanced automotive systems, process control, energy conservation, distributed robotics, weapons systems, manufacturing, distributed sensing command and control, critical infrastructure, smart structures, bio-systems, and communications systems. In order to improve these services to better comply with human needs, the CPS will need to acknowledge the influence of the user or the operator, through Human-in-the-Loop (HiL) controls that take into consideration human intents, psychological states, emotions and actions. As these completely incompatible worlds have to be integrated, innovative solutions are required. In this chapter we provide an overview of the current challenges in designing a smart Vehicular Cyber-Physical System (VCPS) with HiL. We consider the challenges to be of two types or categories, the first related to the cyber-physical part and the second to the human factor in the loop of the system. Nevertheless that the direction is towards development from semi-autonomous to fully autonomous cars, the human driver and/or passenger, and even bystanders play a central role in VCPS performance. We focus on the challenges taking into account the perspectives of the integration in the loop of the VCPS the influence

*Breakthroughs in Smart City Implementation,* 165–188.

of physical processes related to the human behaviour and human emotional states. A special attention is given to the human element in the feedback, as it adds a new dimension of uncertainty in the VCPS and discusses the modelling of driver's behaviour based on his emotional states. The major goal is the identification of areas for further investigation in the scientific area related to CPS with HiL and to outline concrete challenges for future research.

**Keywords:** Cyber physical systems, Human in the loop; Self-coordination, Self-adaptation, Smart Autonomous Vehicle System, Emotion Awareness.

## 7.1 Introduction

Computing and communication capabilities will soon be embedded in all types of everyday objects in the physical environment that surround us. Applications with enormous societal and economic impact will be created by harnessing the capabilities of these web-connected devices for good or bad. Systems that, "bridge the cyber-world of computing and communications with the physical world are referred to as Cyber-Physical Systems (CPSs)" [1]. CPSs are mainly related with the control and monitoring of physical environments and phenomena through sensing and actuation systems consisting of distributed computing and communicating devices [2]. CPSs are systems whose operations are monitored, coordinated, controlled and integrated by a computing and communication core. Due to these facts in the last few years they have attracted significant interest and are being considered to be a frontier scientific research field, which will be of major importance for the years to come as it was mentioned in a report of the European Commission [3]. According to this report a key aspect that is currently missing is the consideration of the socio space, cyber space and physical space interacting at the same time.

The core philosophies of CPSs and the Internet of Things (IoT) are very similar in the field of intensive information processing, comprehensive intelligent services and efficient communications [4]. But CPSs are interconnected systems of collaborating heterogeneous units and components that are envisioned to provide integration of communication, computation and also physical processes. The potential of CPSs is boosted by several recent trends in wireless communications; low cost, low-power, high-capacity, small form-factor computing devices and increased-capability sensors, continuing improvements in energy distribution, alternative energy sources and energy harvesting [5]. CPSs can bring together the discrete and powerful logic of computing and communications, to monitor and control the continuous dynamics of physical and engineered systems. As already described in [6] CPSs must cope with the complexity in the physical environment, together with the lack of perfect synchrony across time and space.

CPSs operate in dynamic contexts and thus have to handle uncertainty that results from phenomena such as interference and noise, abnormal behaviour, rare events, evolving structure, etc. This uncertainty takes another dimension when humans are included in the loop of the CPS. In addition CPSs have to cope with noisy and heterogeneous data [7], to offer robust performance over often unreliable wireless and open communication networks, to operate safely, securely, and efficiently and all this in real-time as they interact with the physical world.

All these different properties and characteristics of CPSs enable new opportunities, pose new research challenges and call for the creation of innovative scientific foundations and engineering principles for CPSs. New approaches based on statistical analysis and mathematics must replace inefficient and testing-intensive techniques. The coupling between the cyber and physical contexts will be driven by new demands and applications. Novel interactions among communications, computing and control must be analysed and understood, and based on this new methods and algorithms that explicitly address the interaction between the physical and cyber subsystems must be developed. They must be based on the integration of the theories of communication systems and computing, sensing and control of physical systems and the interaction between humans and CPSs. The practical implementation of CPSs requires interdisciplinary expertise and skills in communication, and data processing, smart devices and services, security and privacy, mathematical abstractions (algorithms, processes, etc.).

CPSs will transform how humans interact with and control the physical and cyber worlds. The application domains of CPSs involve, but are not limited to, medical systems, assisted living, traffic control and safety, advanced automotive systems, process control, energy conservation, distributed robotics, weapons systems, manufacturing, distributed sensing command and control, critical infrastructure, smart structures, bio-systems, and communications systems [8]. Practical examples of CPS nowadays include different types of medical devices, aerospace systems, intelligent transportation and smart vehicles, defense systems, robotic systems, factory automation, building and environmental control and smart homes and smart cities. Some of the many societal benefits that CPS will deliver will be in the domain of zero-energy buildings and cities, extreme-yield agriculture, near-zero automotive fatalities, perpetual life assistants, and location-independent access to medical care, situation-aware physical critical infrastructure, blackout-free electricity, and safe evacuation from hazardous areas.

This chapter provides an overview of the current challenges in the design of CPS with human in the loop with a focus on smart transport and more precisely on autonomous vehicles. We will focus on the challenges taking into account the perspectives of the physical processes related to the human behaviour, computation and integration in CPS.

The chapter presents knowledge about how some of these challenges are currently handled, what problems have been tackled, which methods have been used to solve them, and how solutions have been evaluated. These insights will help identify areas for further investigation in this particular scientific area and outline concrete challenges for future research concerning the self-adaptation and coordination in CPSs with human in the loop (HiL).

This Chapter is organized as follows. Section 7.2 section describes in detail the main challenges for CPS with HiL, Section 7.3 introduces the smart autonomous vehicles with human in the loop including the current state of modelling driver's behaviour based on emotional states. The final section summarizes the key points of discussion and concludes the Chapter.

## 7.2 Challenges in CPS with HiL

CPSs with HiL are highly integrated, complex and advanced systems that involve the inter-action of many technologies such as wireless sensor networks or robotics, to achieve the sensing, control and environmental adaptation, bringing up many new design challenges. One specific example of CPSs with HiL is future smart autonomous cars. In this case, it is important to note that such highly technical vehicles of the future will still be only machines that will have to learn and understand the context of the world they operate in, thus human context in such systems will become increasingly more important. From one side, proper human engagement in the loop of such CPSs will be necessary, as the autonomous vehicles will continue to be apt to misinterpret the operating environment. From the other side, operators and passengers, as well as bystanders, will influence their characteristics, overall operation, performance and limitations. Thus the proper estimation of the relations between the cyber-physical and human part of the vehicles will be very important for a safe and efficient performance, with the consideration that such relations will be more complex, sophisticated and difficult to be properly modeled and analyzed. A conceptual model of a smart Autonomous Vehicle System (AVS) as a CPS with HiL related to is illustrated on Figure 7.1.

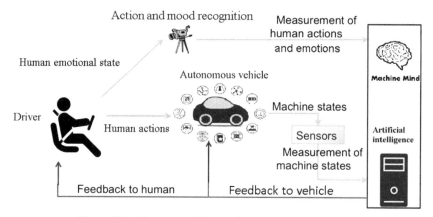

**Figure 7.1** Conceptual Model of AVS as a CPS with HiL.

Such a system, often referred as a type of Vehicular CPS (VCPS), is typically designed as a network of interacting elements with feedbacks instead of being simply a combination of standalone devices. The inherent heterogeneity and integration of different components (sensors, actuators, communications, etc.) pose new challenges to traditional data analysis, communication, control, software theories and artificial intelligence [9]. This often makes system design analysis and modeling inefficient with current approaches. A major difference between a CPS and a typical control system or an embedded system is the use of information exchange and communications, which adds re-configurability and scalability, as well as complexity and potential instability. Furthermore, intelligent sensors and actuators are part of each CPS, together with requirements related to substantially, stricter performance and energy constraints, which are critical for its efficiency and lifetime.

On the other hand complexity lies at multiple temporal and spatial scales, as cyber capabilities and intelligence are embedded in every physical process and component, networking is employed at multiple scales and high heterogeneity is seen across devices and protocols [10]. Timing and spatial precision, uninterrupted connectivity, predictability and repeatability are extremely critical for CPS [11]. In addition to this in many future applications of CPS the human factors will play a major role in their performance, or as shown in Figure 7.1 the humans are in the feedback loop of the system. As very well explained in [12], the scientific community needs to rethink and develop new abstractions and theoretical foundations to allow the proper design, modelling and analysis of the performance of CPS especially with such where the human element is part of the system.

Considering the conceptual model from Figure 7.1 there are some major challenges in the future design, development and operation of a CPS with HiL that can be recognized. We consider these challenges to be of two types or categories, the first related to the cyber-physical part (cyber subsystem and physical subsystem) and the second to the human factor in the loop of the system. In relation to the first, challenging is the implementation of such features as self-monitoring, self-diagnosis, predictability and adaptation in the CPS. For the second type the challenges are connected to modeling and analysis of the influence of the human element in the feedback adding a new dimension of uncertainty in the system.

In many CPS current research initiatives focus is on self-monitoring, self-diagnosis and adaptation to maintain both operability and safety, while also taking into account HiL for system operation and decision making too [13–16]. Typical goals of such self-diagnosis approaches are the detection and isolation of faults, identifying and analyzing effects of degradation, providing fault adaptive control, and optimizing energy consumption [17]. So far, the majority of projects and papers for analysis and diagnosis rely on manually-created diagnosis models of the system's physics and operations [18, 19]. However, the last years have clearly shown that such models are rarely available for complex CPSs and when they do exist, they are often incomplete and sometimes inaccurate, and it is hard to maintain the effectiveness of these models during a system's life-cycle. This is one of the current major shortcomings as the current modeling approaches do not allow a proper and precise design of optimal and fault-free CPSs. A promising alternative is the use of data-driven approaches, where monitoring and diagnosis knowledge can be learned by observing and analyzing system behavior. Such approaches have only recently become possible as CPSs now collect and communicate large amounts of data [20]. This large amount of data can be exploited for the purpose of detecting and analyzing performance. The vision is developing CPSs that can observe their own behavior, recognize unusual situations during operations and inform experts and/or control systems, which can then update operations procedures, and also inform operators or the HiL, who use this information to modify operations or plan for repair and maintenance. The CPS need to adapt and remain operable even in restrictive and hostile conditions, such as network unavailability, hardware failures, resource scarcity, etc. The biggest problem is that an exhaustive catalog of configurations and situations at design time is not a viable solution in the domain of CPS, to model all unanticipated situations in the physical environment. This is followed by the fact that self-adaptive CPS need to base their adaptation actions not only on the current situation, but also to learn from previous situations and improve its performance [21]. A very extensive review on self-adaptation in CPS is done in [22] and a conclusion was that the primary concerns of adaptation in CPS are performance, flexibility, and reliability. The authors' findings show that adaptation in CPS is a cross-layer concern of the technology stack and the main future challenges are how to map concerns to layers and adaptation mechanisms, how to coordinate adaptation mechanisms within and across layers, and how to ensure system-wide consistency of adaptation. In [23] the authors discuss in detail the aspect of self-coordination in CPS and argue that to bridge the gap between the embedded systems with their deterministic behavior and the globally available data and services with their highly probabilistic behavior a certain degree of self-organization together with cognitive capabilities and intelligence implemented in the CPS is crucial.

Since computing and networking systems in CPSs interact with the physical world, predictability (or timeliness) of these systems is an important property that should also be provided. In CPS, temporal predictability of the systems is also an important issue for overall system-wide stability, performance, and safety. In many applications of CPS achieving such real-time properties is much harder as the sensors and connections are wireless and factors such as interference between nodes, dynamically changing network topology, power, etc. are limitations to the practical realization of effective and reliable performing systems. In addition, it is worth mentioning that new security challenges are introduced by CPSs, especially with HiL, challenges that cannot be fully handled via traditional security and/or cryptographic means. In many CPS the devices, such as sensors, have limited capabilities and are not controlled by a central control center; thus, the implementation of computationally expensive security mechanisms can be challenging especially with and increased complexity when human factors are part of the system.

The self-adaptation, self-coordination, connectivity together with the predictability issues should be redefined in the framework of CPSs with HiL. Such systems will usually consist of both sensor and actuation networks with different capacities and reliability, but the human presence in the loop will add another dimension of uncertainty. To model such heterogeneous system, is a very big challenge as an innovative scientific and technological turn is necessary. To compromise all those critical aspects of CPSs with HiL, sophisticated signal and data processing techniques, estimation and analysis of human behavior and influence, coupled with novel network, connectivity and security approaches, should be researched and designed to provide highest performance and efficiency levels for such CPS.

In CPSs the human context will become increasingly more important as most future technologies will be human-aware. In a future cyber-physical world, with smart cities, smart factories, smart vehicles, etc., humans will be relieved in many of their tasks, but such smart cyber-physical environments will also require humans to be more responsible and vigilant as they will exercise much higher levels of cognition in interpreting and reacting to the decisions that the CPSs will be making. For example, as mentioned before, smart VCPS will continue for a long time in the future to misinterpret the world in which they operate as they are far away from the point to handle things and random occurrences we humans deal with all the time, such as rapidly changing weather conditions, complex terrain, unusual objects in the road, etc., and the unpredictable nature of human behaviour. Thus proper human engagement with the cyber-physical world will continue to be a critical area of development for VCPS.

Future CPSs will most likely bolster a much stronger tie between humans and the control loops of the system. A CPS with HiL is a system that takes human response into consideration and human presence and behavior becomes a key part of the system instead. In this case it is essential to develop and integrate reliable and accurate human behavior modeling techniques that attempt to learn and predict human behavior. Capturing human behavior by extending system identification or other modeling techniques is extremely difficult due to complex physiological, psychological, emotional and behavioral aspects of human beings. Also, the level of modeling depends on application requirements. Although requirements are different for different applications, a significant portion of HiL applications have to address some common challenges, e.g., user specific metrics, thresholds and parameters, estimation of the change of human behavior over time, and required sensing technology to sense the appropriate aspects of human behavior. Human behavior must be modeled for large number of applications before general principles and theories emerge to address these issues. Clustering, data mining, specialized models based on human physiology and behaviors may all be techniques to be enhanced before applied in the development of CPS with HiL. Robust CPS systems will likely require predictive models to avoid problems before they occur; consequently advances to stochastic model predictive control are also required. It is also unlikely that any models developed initially to design the controllers will remain accurate as the system and human behaviors evolve over time. Hence, cognitive, self-learning and adaptive features in the control loop of the CPS with HiL will be necessary to be implemented.

Currently, the techniques that model certain aspect of human behavior are either very general or very specific. For example, Smart Thermostat [24] uses a Hidden Markov Model (HMM) to model occupancy and sleep patterns of the residents in a home to save energy, which captures human behavior from a very high level. Another example, describes a new paradigm called Body Coupled Communications (BCC) for Wireless Body Area Network (WBAN) that leverages the human body as a communication channel [25]. In such application, sensors are implanted in a human body that are capable of monitoring a wide range of physiological and emotional states and the human serves as a communication channel to transmit the sampled data to a centralized monitoring entity.

Another challenge is related to determining how to incorporate human behavior models into the formal methodology of description of CPSs with HiL. Human behaviour and the detection of human emotions can lead to emotionally-aware applications, or applications meant to react to natural human behaviour with the goal of using this information to help improving CPSs performance. Recently, such systems are referred as HiL emotionally-aware CPSs, in which human nature and emotions are responsible for direct system actuation and affect the system's actions. For such CPSs the derivation of advanced mathematical models or machine learning techniques that can reliably classify and possibly predict human nature and behaviour is a big challenge [26]. Moreover, even if we have a model of human behavior, it is not clear where to place the model for "each" specific application and to incorporate the human behavior as part of the system itself. Further below we consider one of the applications of CPSs with HiL in Smart Environments – the Vehicular Cyber-Physical Systems Repetition where human factor is inherent part of the system. We bring forward to the attention of the reader the problem with the recognition and modeling behavior and specifically emotional conditions of the human for the application in HiL emotionally-aware VCPS.

## 7.3 Smart Autonomous Vehicles with Human in the Loop

### 7.3.1 Vehicular Cyber-Physical Systems

CPSs with application in transport are rapidly advancing due to progress in real-time computing, control and artificial intelligence [27]. Much research has been conducted in the field of mobility and traffic management, routing and data dissemination, aggregation, and safety. In the recent years VCPSs have attracted a significant interest offering different applications mostly in the aspect of smart cities ensuring safer transportation. Besides their traditional functionality automobiles today are being integrated with more and more sensors and actuators and, they already possess intelligence with respect to the environment and the driver, which make them typical VCPS. Autonomous vehicles are reality. Many countries are elaborating laws and legislation allowing autonomous vehicles to operate on the roads together with human driven vehicles. In a report filed with the California Department of Motor Vehicles, Waymo subsidiary of Google reports that in 2016 its self-driving cars clocked 635,868 miles (1,023,330 km), and required human intervention 124 times. That is one intervention about every 5,000 miles (8,047 km) of autonomous driving. The more impressive being the progress it's made in just a single year: human interventions fell from 0.8 times per thousand miles to 0.2, which translates into a 400% improvement. With such progress, Google's cars will easily surpass the driving ability of a standard human driver.

All contemporary cars are sold with a good amount of smart components. Such components include engine control sensors, comfort and safety features. Although these intelligent components help to contribute to safe mobility, they lack human centric characteristics, such as personalization, or knowledge about the user in order to achieve save and secure autonomous driving. A human-centric vehicular system uses smart sensors to gather information, make fast decisions and adapt its behaviour, leaving the human free to make higher level decisions. Currently there is quite a progress in autonomous driving [28] where the human centric element is a primary concern and various threats and vulnerabilities have been discovered from different models of smart cars [29]. But there are still many problems to be solved related to modelling the interaction and dynamics of the cyber and physical components; analysis of interdependencies to enable secure and robust operation and security vulnerabilities with domain specific knowledge and how to model and use the "human-in-the-loop" to ensure secure and save transportation. VCPS where the human becomes an integral part of the control-loop and his nature directly affect the system's actions. For instance, when a person is stressed, nervous or in a very bad emotional condition or mood, the system can respond and thus, the person's emotional state will be responsible for direct system actuation.

Practically, VCPS are CPS with the human element in the loop as the driver in many scenarios acts as a feedback control unit for all generated information. In VCPS with HiL, the human factor is of major importance as the driver's perceptions/reactions and emotional states are fundamental to design of safety VCPS. However, most of the existing works have not studied this problem looking at both the VCPS and the influence of the emotional state of the human being a part of the system, or being in the feedback of the loop. Building secure VCPS is a non-trivial task, especially when analysing how the driver reacts on the VCPS environment, as a whole this is a system dependent on the cyber and physical environment as well as on human related negative factors, such as information overload, confusion, distraction and different states of emotion. Major issues in the design of such VCPS with HiL are related to the development of more intelligent systems ensuring reliable performance of the VCPS in heterogeneous environments. For this to be achieved it is necessary the whole system to be able to make correct, informed decisions, based on the analysis and prediction of the behaviour or physical and emotional state of the driver in order to guarantee safe, reliable and correct interaction between him and the CPS environment in different situations.

## 7.3.2 Modelling Emotionally Aware Vehicles

When it comes to road security, factors such as driver's fatigue, speeding and a general reckless behaviour impacts the driver's safety. Another large factor is the emotional state of the person behind the wheel for example – road rage and other non-explicit emotional responses the human driver experiences behind the wheel that can have an impact on the way he drives. A very basic example is that if the driver feels agitated or angry, he or she may end up driving more aggressively and dangerously. Considering cars are becoming much smarter, with semi-automated functions (e.g. collision detection, automatic parking, road signs recognition etc.) or even completely autonomous (Google's self-driving cars), vehicles will include technology that can detect when the drivers are tired or emotionally frustrated so the vehicle can respond and modify its behaviour based on the emotional state of the driver or the passengers. According to many recent research articles there are several applications of emotional intelligence in smart vehicles:

- Increased Road Safety. With emotion enabled vehicles, driver assisted technologies can monitor for driver attention, identifying fatigue, distraction and frustration, rage or stress to prevent accidents before they happen. With the help of facial expression recognition, patterns in the gathered data emerge that can monitor, define, and predict driver's and other human behaviour that can influence vehicle operation as mentioned in 7.2

- Augmented Operation of Autonomous Cars. In self-driving cars, emotional intelligence can help control the vehicle's operation and help determine in real time when to introduce hints and narrative in order to counter perceived emotional stress of passengers and help them relax or offer appropriate in-car entertainment.
- Best In-Car Experience. With emotion awareness, the vehicle's smart systems can make highly personalized recommendations and adjustments based on the mood of the car's occupants, such as alternate routes, interesting stops along the road, audio system, or climate control.
- Ultimate Personalization. Collecting emotional data to develop a driver's unique profile can allow for complete customization of how vehicles operate, and will greatly influence the experience both drivers and passengers will have.

To be able to develop these applications a universal driver's behaviour model is needed. According to Vaa [30] the difficulty in modelling driver's behaviour resides in the lack of a thorough and comprehensive understanding of human cognition and emotion, i.e. how drivers think and feel, consciously, pre-consciously and unconsciously. The existing models address different aspects of driver's behaviour such as cognitive distraction [31], fatigue detection [32], emotional stress [33], routine variations [34], the brake/throttle pedal and the steering wheel manipulation [35]. Each of the aspects has its significance, but there is no general integration that links the diversity of aspects into a complete and all-embracing driver's behaviour model.

How an emotion can be modelled and described with quantitative characteristics? As a result of improvements in camera technology, processor chips, machine learning and computer vision algorithms, machines are poised to take a big leap in their ability to understand humans from their facial expressions, the movement of the eyes and the head, gestures, gait and posture. On Figure 7.2 is illustrated a generalized VCPS conceptual model taking into account the relevant to driving emotional and cognitive aspects: awareness, fatigue, drowsiness and mood.

The dangers of drowsy driving have been widely documented. A very recent study found that car crashes become more likely in the few days after adjusting the clocks for daylight saving time and people lose an hour of sleep [36]. There are no lawful ways to prevent people from driving while sleep deprived because there is no benchmark for sleepiness like there is for drunkenness – even though the effects of sleep deprivation are equivalent to those of being legally drunk.

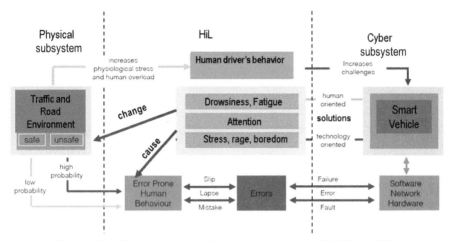

**Figure 7.2** Conceptual model of an emotionally aware VCPS with HiL.

Driver's drowsiness, fatigue and attention can be measured in a non-obtrusive way from camera based on facial movements such as excessive blinking, yawning, head movement and gaze tracking. To detect such conditions we have developed a framework using an active multi-camera system in 3D space for human gaze tracking and face orientation determination combined with 3D face and emotion recognition based on multiple kernel learning [37]. We employ a fixed camera in order to determine the position of the human face and its features, most importantly the eyes. The block diagram shown in Figure 7.3 describes the main stages and their consistency for processing the information until the final aim is achieved. By means of the Supervised Descent Method (SDM) for minimizing a nonlinear least squares function; we can compute correctly the position of the two eyes using 6 landmarks for each eye and the position of the head. Then, an active pan–tilt camera is oriented to one of the users' eyes, and in this way, a high-precision gaze direction determination is accomplished, thus being able to deter-mine the focus of attention of the user [38].

Using only camera facial emotions of the driver such as anger, stress and boredom can be recognized. Facial expressions are universal and independent of race, culture, ethnicity, nationality, gender, age, religion, or any other demographic variable. All people express emotions on their faces in exactly the same way. Many research efforts are concentrated in the task of facial expression recognition as this can be seen in the multiple published papers in recent years [39]. According to [40] facial expressions (i.e., contractions of facial muscles) induce movements of the facial skin, also changes in the

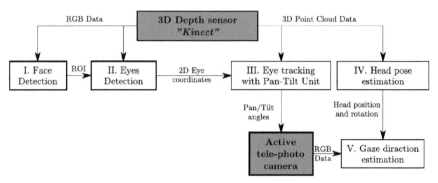

**Figure 7.3**    Block diagram of the proposed system for gaze and head movement tracking for awareness detection.

position and appearance of facial components (e.g., mouth corners are pulled up in smile) and changes in facial transient features (e.g., crow-feet wrinkles around the eyes deepen in genuine smile).

One approach to reduce uncertainty brought by the human in the loop in the VCPS is to recognize emotional state of the driver. By monitoring the facial expressions the VCPS can build patterns in the data that will help define and/or predict the driver's behaviour and his reactions to abnormal situations occurring unexpectedly. Different techniques for expression recognition can be used to achieve this goal. In [41] we have developed a system based on formation of feature vectors by concatenating the landmarks data from Supervised Descent Method, applying Principal Component Analysis (PCA) and use of these data as an input to Support Vector Machine (SVM) classifier. The experimental results show improvement of the recognition rate in comparison to some state-of-the-art facial expression recognition techniques. In [42] we concentrated our efforts in stress and anxiety recognition from facial images based on Active Shape Models with 2D profiles and classification with kernel SVM. One of the most effective methods for expression recognition was presented in [43] where we propose an algorithm for Action Unit recognition from still images. The algorithm utilizes contemporary feature selection approach based on analysis of the data graph. The feature selection is fed by features extracted using a state of the art algorithm for facial key points detection.

At this stage there are very few problems left with automatic facial expression recognition using only information from cameras i.e. the sensing conditions are more or less standardised, nearly frontal faces, constant/acceptable illumination conditions. In [44] the authors ask the question: "Is automatic facial expression recognition of emotions coming to a dead end?" and they give a very accurate description of the current state of this field where more and more elaborate and complex machine learning and computer vision approaches are proposed without taking into account the shift from the personal computer to the portable technology. The future of facial expression recognition is the multimodal emotion sensing. An accurate and real-time emotion recognition system cannot be based only on images and video. It is important to use all the available modalities: voice, touch, bio signals etc. But we must take into account that the smart vehicle will be connected to all our personal and wearable smart devices so biometric data can be shared to assure best possible estimation of the influence of emotional states of the HiL in the VCPS for correct decision and best possible outcome in any situation.

## 7.4 Conclusion

Many automotive companies are heading towards implementing solutions based on emotion recognition for semi-autonomous and fully autonomous vehicles. They are trying to solve a couple of problems related to the user's acceptance of the semi-autonomous and fully autonomous vehicles, user behaviour, sudden changes in environmental conditions etc., and causing a mistrust in the whole system's performance. The reasons of mistrust are mainly based the on human factors such as mood, fear and insecurity [45]. The users of the autonomous cars see and feel the robotic-like drive of the vehicle that follows planned trajectories. This behavior often feels unnatural compared to human driving. The users feel insecure if the autonomous vehicle recognizes and evaluates traffic situation correctly or if a critical maneuver has to be performed. These two shortcomings can be diminished when implementing emotion recognition into the vehicle. The recognition of emotional state of the user will allow adaptation of the vehicle's dynamics to the current emotion such as joy, resentment or fear and furthermore can make the driving behavior much more transparent to the users in a natural way and customized according the his desires. Carmakers are testing dashboard-mounted sensors that watch drivers for signs of fatigue or distraction and even have already offered a driver attention and drowsiness systems build in their car models that use sensors and cameras to monitor a driver's steering habits, ability to remain within a lane, and time at the wheel. These new technologies have the potential to make driving safer by watching the driver's behavior and implementing attention or mood-altering solutions. They are also testament that technologies are everywhere-watching, using users' faces and emotions as content to determine, and then change their behavior.

In this chapter we have considered some of the current challenges in designing a smart VCPS with HiL. Taking into account a general conceptual model some major challenges in the future design, development and operation of VCPS with HiL were recognized. We classify them in two categories: the first related to the cyber-physical part and the second to the human factor and human behavior in the loop of the system. We focus on the challenges related to determining how to incorporate human behavior models into the formal methodology of description of VCPS, or more specifically how to incorporate the human behavior as part of the system itself. We examine a generalized VCPS model taking with the relevant to driving emotional and cognitive aspects: awareness, fatigue, drowsiness and emotional state and propose some approaches of their estimation, for the goal of reducing the uncertainty brought by the HiL in a VCPS.

The overall impact of the CPS on the human society is currently largely unknown, since social sciences can hardly cope with the speed of introduction of new technologies. But one important question arises about CPS with HiL: Whose Responsibility: Human vs. CPS? How much do users merge with CPS? Are they partially integrated in the existing system or do they fully depend on the advanced conveniences offered by the CPS? For example, when an emergency happens on a freeway, would systems release the control of automobiles to drivers or not? If not, would drivers allow their lives to be in the sole control of machines, argue the authors in [46] and [47]. And in the case of some damage, who or what is liable for the damage. Another challenge that arises is the trust or social acceptability of autonomous vehicles [48]. It should also be possible to interpret human trust in an autonomous vehicle agent based the emotional state of the human. The lack of answers to these legal liability issues could actually prevent useful CPS technology from being implemented.

## Acknowledgements

This work was supported by the contract No DN 07/22 15.12.2016 "Self-coordinating and Adaptive Wireless Cyber Physical Systems with Human in the Loop" funded by the National fund for scientific research, Bulgaria, 2016–2018.

# References

[1] Rajkumar, R., Lee, I., Sha, L., and Stankovic, J., (2010). "Cyber-physical systems: the next computing revolution", in *Proceedings of the Design Automation Conference (DAC), 2010 47th ACM/IEEE*, Anaheim, CA.

[2] Lee, E. A. (207). *Computing Foundations and Practice for Cyber-Physical Systems: a Preliminary Report*. Technical Report UCB/EECS-2007-72, Berkeley, CA: University of California.

[3] Cyber-Physical Systems (2013). "Uplifting Europe's innovation capacity," in *Proceedings of the Report from the Workshop on Cyber-Physical Systems: Uplifting Europe's Innovation Capacity, 29th and 30th October 2013*, Brussels.

[4] Wan, J., Chen, M., Xia, F., Di, L., and Zhou, K. (2013). From machine-to machine communications towards cyber-physical systems. *Comput. Sci. Informat. Syst.* 10, 1105–1128.

[5] Jabeur, N., Sahli, N., and Zeadally, S. (2015). Enabling cyber physical systems with wireless sensor networking technologies, multiagent system paradigm, and natural ecosystems. *J. Mob. Informat. Syst.* 2015, 15.

[6] Lee, E. A., and Cheng, A M. K. (2015). the past, present and future of cyber-physical systems: a focus on models. *Sensors (Basel)* 15, 4837–4869.

[7] Rajhans, A., Bhave, A., Ruchkin, I., Krogh, B. H., Garlan, D., Platzer, A., et al. (2014). Supporting heterogeneity in cyber-physical systems architectures. IEEE Trans. Automatic Control, 59, 3178–3193.

[8] Geisberger, E., and Broy, M. (2015). *Living in a Networked World: Integrated Research Agenda Cyber-Physical Systems (agendaCPS)*. München: Herbert Utz Verlag.

[9] Tzagkarakis, G., Tsagkatakis, G., Alonso, D., Asensio, C., Celada, E., Panousopoulou, A., et al. (2015). *Signal and Data Pro-cessing Techniques for Industrial Cyber-Physical Systems Cyber-Physical Systems From Theory to Practice*. Boca Raton, FL: CRC Press, 181–226.

[10] Cao, X., Cheng, P., Chen, J., and Sun, Y. (2013). An online optimization approach for control and communication codesign in networked cyber-physical systems. *Ind. Inform. IEEE Trans.* 9, 439–450.

[11] Kim, K.-D., and Kumar, P. (2012). "Cyber physical systems: a perspective at the centennial", in *Proceedings of the IEEE*, Piscataway, NJ, 1287–1308.

[12] Poovendran, R. (2010). Cyber-physical systems: close encounters between two parallel worlds [point of view]. *Proc. IEEE* 98, 1363–1366.

[13] Lee, E.A. (2008). "Cyber physical systems: design challenges," in *Proceedings of the Object Oriented Real-Time Distributed Computing (ISORC) 11th IEEE International Symposium Orlondo*, FL, 363–369.

[14] Rajkumar, R., Lee, I., Sha, L., and Stankovic, J. (2010). "Cyber-physical systems: the next computing revolution", in *Proceedings of the 47th Design Automation Conference, DAC* New York, NY: ACM, 731–736.

[15] Evans, P. C., and Annunziata, M. (2012). *Industrial Internet: Pushing the Boundaries of Minds and Machines"*, Technical report, General Electric, Oak Ridge, TN.

[16] Forschungs union (2013). *Promotorengruppe Kommunikation. Im fokus Das Industrieprojekt Industrie 4.0, Handlungsempfehlungen zur Umsetzung*. (New York, NY: Forschungsunion Wirtschaft-Wissenschaft).

[17] Christiansen, L., Fay, A., Opgenoorth, B., and Neidig, J. (2011). "Improved diagnosis by combining structural and process knowledge," in *Proceedings of the Emerging Technologies Factory Automation (ETFA), 2011 IEEE 16th Conference, Bilbao.*

[18] de Kleer, J., Janssen, B., Bobrow, D. G., Saha, T. K., Moore, N. R., and Sutharshana, S. (2013). "Fault augmented modelica models", in *Proceedings of the The 24th International Workshop on Principles of Diagnosis*, Portland, OR, 71–78.

[19] Klar, D., Huhn, M., and Gruhser, J. (2011). "Symptom propagation and transformation analysis: A pragmatic model for system-level diagnosis of large automation systems", In *Proceedings of the Emerging Technologies Factory Automation (ETFA), 2011 IEEE 16th Conference*, Barcelona, 1–9.

[20] General Electric. (2012). *The Rise of Big Data – Leveraging Large Time Series Data Sets to Drive in-Novation, Competitiveness and Growth – Capitalizing on the Big Data Oppurtunity*. Technical report, General Electric Intelligent Platforms.

[21] Gerostathopoulos, I., Bures, T., Hnetynka, P., Keznikl, J., Kit, M., Plasil, F., et al. (2016). Self-adaptation in software-intensive cyber-physical systems: from system goals to architecture configurations. *J. Syst. Softw.* 122, 378–397. doi: 10.1016/j.jss.2016.02.028.

[22] Muccini, H., Sharaf, M., and Weyns, D. (2016). "Self-adaptation for cyber-physical systems: a systematic literature review", in *Proceedings of the SEAMS '16 Proceedings of the 11th International Symposium on*

*Software Engineering for Adaptive and Self-Managing Systems, Austin, Tx*, 75–81.

[23] Rammig, F. J. (2012). "Self-coordination as fundamental concept for cyber", in *Proceedings of the 2nd SBC Workshop on Autonomic Distributed Systems – WoSiDA 2012*, Ouro Preto, 45–48.

[24] Lu, J., Sookoor, T., et al. (2010). The smart thermostat: using occupancy sensors to save energy in homes". in *Proceedings of the 8th ACM Conference on Embedded Networked Sensor Systems SenSys*, Zurich.

[25] Schirner, G., et al. (2013). The future of human-in-the-loop cyber-physical systems. *Computer* 1, 36–45.

[26] Nunes, D., Zhang, P., and Sa Silva, J. (2015). A survey on Human-in-the-Loop applications towards an Internet of All. *IEEE Commun. Surv. Tutor.* 17, 2015.

[27] Bradley, J. M., and Atkins, E. M. (2015). Optimization and control of cyber-physical vehicle systems. *Sensors (Basel).* 15, 23020–23049. doi: 10.3390/s150923020, 2015.

[28] Okuda, R., Kajiwara, Y., and Terashima, K. (2014). "A survey of technical trend of ADAS and autonomous driving". in *Proceedings of Technical Program 2014 International Symposium on VLSI Technology, Systems and Application (VLSI-TSA)*, Hsinchu, 1–4.

[29] Humayed, A., Luo, B. (2015). "Cyber-physical security for smart cars: tax-onomy of vulnerabilities, threats, and attacks", in *Proceedings of the ICCPS '15 ACM/IEEE Sixth International Conference on Cyber-Physical Systems*, Stockholm, 252–253.

[30] Vaa, T. (2013). "Proposing a risk monitor model based on emotions and feelings: Exploring the boundaries of perception and learning", in *Book: Driver Dis-Traction and Inattention: Advances in Research and Countermeasures*, Vol. 1, eds M. A. Regan, J. D. Lee, and T. W. Victor (Farnham: Ashgate), 103–119.

[31] Yuce, A., Gao, H., Cuendet, G. L., and Thiran, J.-P. (2017). "Action units and their cross-correlations for prediction of cognitive load during driving", in *Proceedings of the IEEE Transactions on Affective Computing*, London, doi: 10.1109/TAFFC.2016.2584042

[32] Tu, W., Wei, L., Hu, W., Sheng, Z., Nicanfar, H., Hu, X., et al. (2016). "A survey on mobile sensing based mood-fatigue detection for drivers," *Smart City 360°*, eds A. Leon-Garcia, R. Lenort, D. Holman, D. Stas, V. Krutilova, P. Wicher et al. (Cham: Springer), 3–15. doi: 10.1007/978-3-319-33681-7-1

[33] Gao, H., Yuce, A., and Thiran, J.-P. (2014). "Detecting emotional stress from facial expressions for driving safety," in *Image Processing (ICIP), 2014 IEEE International Conference*, Paris, doi: 10.1109/ICIP.2014.7026203

[34] Banovic, N., Buzali, T., Chevalier, F., Mankoff, J., and Dey, A. K. (2016). "Modeling and understanding human routine behavior," in *CHI '16 Proceedings of the CHI Conference on Human Factors in Computing Systems*, San Jose, CA, 248–260.

[35] Wang, W., Xi, J., and Chen, H. (2014). *Modeling and recognizing driver behavior based on driving data: a survey*. Cairo: Hindawi Publishing Corporation, doi: 10.1155/2014/245641.

[36] Smith, A. C. (2016). Spring forward at your own risk: daylight saving time and fatal vehicle crashes. *Am. Econ. J.* 8, 65–91.

[37] Tonchev, K., Panev, S., Manolova, A., Neshov, N., Boumbarov, O., and Poulkov, V. (2015). Gaze tracking, facial orientation determination, face and emotion recognition in 3d space for neurorehabilitation applications. *Neuro Rehabil. Brain Int.* 40, 51–89.

[38] Manolova, A., Panev, S., Tonchev, K. (2014). "Human gaze tracking with an active multi-camera system", in *Proceedings of the Biometric Authentication in Lecture Notes in Computer Science, Springer, Revised Selected Papers of First International Workshop BIOMET'2014*, Sofia, 23–24.

[39] Chiarugi, F., Giannakakis, G., and Pediaditis, M. (2014). "In-depth analysis of state-of-the-art for facial expression analysis." 2014, in *Proceedings of the WP 5.1 in SEMEiotic Oriented Technology for Individual's Cardiometabolic Risk Self-Assessment and Self-Monitoring*, Paris.

[40] Pantic, M., and Rothkrantz, L. J. (2004). Facial action recognition for facial expression analysis from static face images. *IEEE Trans. Syst. Man Cyber.* 34, 1449–1461.

[41] Manolova, A., Neshov, N., Panev, S., and Tonchev, K. (2014). "Facial Expression Classi-fication using Supervised Descent Method combined with PCA and SVM", in *Proceedings of the Biometric Authentication in Lecture Notes in Computer Science, Springer, Revised Selected Papers of First International Workshop BIOMET'2014*, Sofia, 23–24.

[42] Penev, M., Manolova, A., and Boumbarov, O. (2014). "Active shape models with 2d profiles for stress/anxiety recognition from face images", in *Proceedings of the 9th International Conference on Communications, Electromagnetics and Medical Applications,* Sofia, 16–18.

[43] Sechkova, T., Tonchev, K., and Manolova, A. (2016) "Action unit recognition in still images using graph-based feature selection", in *Proceedings of the IEEE 8th International Conference on Intelligent Systems (IS)*, Sofia. 4–6.

[44] Gunesa, H., and Hung, H. (2016). Is automatic facial expression recognition of emotions coming to a dead end? The rise of the new kids on the block. *Image Vision Comput.* 55, 6–8.

[45] Kraus, S., Althoff, M., Heißing, B., and Buss, M. (2009). "Cognition and emotion in autonomous cars", In *Proceedings of the IEEE Intelligent Vehicles Symposium*, Gothenburg, 635–640.

[46] Shi, W., and Wang, S. (2008). "Building self-adaptive cyber physical systems using unreliable components", in *Proceedings of the National Workshop for Research on High-Confidence Transportation Cyber-Physical Systems: Automotive, Aviation & Rail*, Washington, DC, 18–20.

[47] Marwedel, P., and Engel, M. (2016). "Cyber-physical systems: opportunities, challenges and (some) solutions", in *Proceedings of the Management of Cyber Physical Objects in the Future Internet of Things Part of the series Internet of Things*, Springer: Berlin, 1–30.

[48] Basu, C., and Singhal, M. (2016). "Trust dynamics in human autonomous vehicle interaction: a review of trust models", in *Proceedings of the AAAI Spring Symposium 2016*, Palo Alto, CA, 21–23.

# Biographies

**Assoc. Prof. Agata Manolova** is with the Faculty of Telecommunications at the Technical University of Sofia (TU-Sofia), Bulgaria. She finished her Ph.D. conjointly between the TU-Sofia and the Joseph Fourier University, Grenoble, France. Her domains of interest are Pattern Recognition, Computer Vision, Statistics, Image and Video processing, Biometrics, Processing of EEG,

EMG signals, Ambient Assisting Living Systems. Dr. Manolova has participated in several scientific projects both national and international. She has published more than 40 papers in conferences and journals. She is laureate of Fulbright scholarship for 2013 working at the Computer Vision Laboratory at the Little Rock University, Arkansas, USA on a project concerning recognition of human emotions. Assoc. prof. Manolova is an IEEE member. She is teaching courses for Bachelor and Master degree students both in English and French language.

**Prof. Vladimir Poulkov** has more than 35 years of industrial, research and teaching experience in the field of telecommunications. His expertise is in the field of information transmission theory, modulation and coding interference suppression, power control and resource management for next generation telecommunications networks, cyber physical systems. He has been leader of many national and international industrial, R&D and educational projects. He is author of more than 150 scientific publications and is leading BSc, MSc and Ph.D. courses in the field of Information Transmission Theory, Broadband Transmission and Access Networks. Currently he is Head of "Teleinfrastructure" and "Electromagnetic Compatibility of Communication Systems" R&D Laboratories at the Technical University of Sofia, Chairman of Bulgarian Cluster Telecommunications, Vice-Chairman of European Telecommunicaitons Standartization Institute (ETSI) General Assembly, Senior IEEE Member.

**Krasimir Tonchev** a senior researcher leading scientific and research activities at the "Teleinfrastructure Lab", Faculty of Telecommunications, Technical University of Sofia, Bulgaria. His research interests include, on the theoretical side, large scale Kernel Machines, modeling of dynamical behavior, Bayesian modeling, and on the application side, 2D and 3D facial analysis for soft biometrics, affective computing and general scene understanding from video. He has lead and coordinated many national and international scientific projects in the field of signal analysis, computer vision, reliable face and emotion recognition, and ambient assisting living systems. He also participated in innovative commercial product development including embedded vision and image processing with a large scale volume production. He has published many scientific papers in the field of facial analysis and general human behavior understanding. He is an IEEE member.

# 8

## What Makes Cities Bloom and Prosper?: Connected & Cooperating People

Jaap van Till

Prof Emeritus Network Architecture and Digital Infrastructures
Tildro Research, Rhenen, The Netherlands

### Abstract

People are amazed at the complexity and size of anthills in Africa, but what they do not realize that these are built not by leadership or architectural instructions, but by emergent behaviour from hundreds of thousands of face-to-face interactions between ants and subsequent massive numbers of small actions, by groups of ants with different functions. From these they together build ventilation channels, humidity control, storage rooms, escape tunnels, carry food, etc etc. These face-to-face interactions are done by smelling each other with only eight different smell codes. In the present cities of the world most tasks are done by also face-to-face in real life (IRL) meetings of people, but our small and big human-hills are much influenced by tele- (on a distance) communication also. By watching TV broadcasts, by talking on the telephones together, by looking and interacting from their PC's, smartphones or Laptops connected to fast Internet. These hundreds of millions of parallel messages and interactions daily have a massive effect on what cities look like and how they evolve.

In this chapter we present a number of ways these interactions influence and exert power on the cities. From e-mail and telephone to more complicated collaboration tools that are carrying information, knowledge and practical knowhow to get things done, to wisdom and conscious shared visions. It is not the ICT technology only, but what individuals and groups of **people do together** with those power tools that matter.

*Breakthroughs in Smart City Implementation,* 189–214.

**Keywords:** Smart Cities, Social Networks, Connectivity, Collaboration, Network Effects, Synergy, Synthetic Apertures, Collective Intelligence, Corridoria, network effects, telescope metaphor, value creation, chains of city regions, global brain, weavelets, commons, synthecracy.

## 8.1 Introduction

An interesting example of a very successful, recently constructed, mega-city is Dubai, an important hub in international airline flights. In a promotional video about this city, Parag Khanna, famous map-maker of international infrastructures [1], tells what is so good about the resilience and strategy of Dubai's architects [2]. But the question remains after you see this film: what is the MOTOR of that city? Therefore, one of the research questions of this book chapter is "How can we make small and big cities bloom? What drives prosperity there?

Many scholars have published answers to these questions. For instance Richard Florida derived guidelines for American cities who noticed that some cities where successful in attracting talented young people who started new businesses. He published the three T's related to 3 magnets: *Technology, Talent, and Tolerance*; which do interact and mix to produce a kind of 'chimney effect' of growth. Cities that do not create these conditions can be shown in the statistics to decline fast. The movers and kickers leave those towns. Other more recent influential work is in the books and lectures of Jeremy Rifkin [3]. His main advice for prosperity and attractiveness of city area's is to construct and improve the three vital infrastructures for (mega-poli) city-area's: Energy distribution infrastructure, ICT & communication digital flows infrastructure, physical goods and persons & goods transport infrastructure for logistics. For these infrastructures in order to operate & maintain there must be in place: a Communications Internet, an Energy Internet and a Transportation Internet!!! A smart energy grid cannot operate without a smart digital network. Same applies for logistics. That can only function without a vast grid of computer and human communications.

I will focus in this chapter not on the vital conditions, but on what people DO TOGETHER in cities based on those infrastructures and why that, as a motor, can create value and wealth.

## 8.2 Human Networking

It should be clear that the activity of "networking" is very much empowered, supported and speeded up in recent decades by Internet-Email and Social

Media use. What are less visible are the informal networks of people who help each other or refer them to others who maybe can solve the stated problem.

There is always (1) a formal network of decision makers with authority and power who delegate tasks in hierarchies and to which you have to report and (2) a second network of specialists who have knowledge on a certain, usually very narrow but deep, subject and its solutions to problems. Problem is that these much respected specialists do not refer to each other because they consider themselves as the centre of the universe and often do not think other people know anything relevant. Fortunately there is also (3) a third network present in every organization, otherwise the organization would not have existed anymore. It is the network of "carriers"/gatekeepers. They keep decision makers and specialists in balance by carrying issues forward, often by transferring questions to somebody somewhere that they think knows everything about a solution, and if not transfer it to somebody else, etcetera. Scientists are very good at this, which can usually solve anything insides or outsides the organisation within six or seven steps. There are a number of rules & ethics, elucidated in my published lecture [4] for this very important social "networking" which can take place on eMail, in corridors, at lunch, in coffeehouses or pubs. These carriers/transfer agents test each other out to know if they can be *trusted* to solve problems. So there is a lot more behind this than shaking hands at parties and exchanging business cards. You must have done work together and shown your abilities to be included in one of the many "networks" in a city. Otherwise you will be bypassed.

Interestingly there is another level of social networking which is less well known but even more important not only for the functioning of cities but for the stability of society. In 1973 the sociologist M. Granovetter noticed that graduates from universities got more successful jobs from referrals & recommendations by distant acquaintances (weak links) than from their own family members (strong links). These 'weak links' are in fact very strong and effective, that is why I have given them another name "Btwieners", see [4]. Key is that they are respected in more than one tribe, like the traveller Gandalf the Gray in Tolkien's books, welcome wherever he arrived. This rare breed of super-networker I call Btwiener who is able to **interconnect people** in different tribes/families to work together in teams based on their respective shown *different* abilities/skills/crafts allowing to contributing whatever their background is. The super-networker activates their ability to learn very fast from their colleagues. Such open teams learn from their clients and from the environment they work in and improve and innovate so fast that others can copy but not overtake them. I call that '**trans tribal collaboration**'.

Prof. Peter Csermely studied [5] these interconnecting people and found that a vast variety of networks, varying from proteins, people, brain repair cells to eco-systems also use weak interactions to function and solve damages. These mobile units stabilize the whole system. And only a few of them are around. You will not find them on any formal organization chart though.

Social networks on Internet support & speed up very much the activities of 'networking' as well as the 'Btwiening'. I recommend that cities identify and support these Btwieners because although they most often do their work voluntary, the effect of what they do is most often underestimated. They are a make or break asset.

## 8.3 The Network Effects and the Value that They Generate

Now, after we have shed some light upon the people who "connect" others, we should focus more on 'why is it valuable when people communicate, connect together and cooperate?'. This brings up the analysis of the so called 'Network Effects' [6], often quoted but seldom well understood.

*"Everything wants to be connected"*, is the famous 'Renan Law' of my friend and colleague Sheldon Renan. But, why does everything in the universe want that? What is this self-interest or incentive to connect? And what does one do after the technical connection? On how many levels and how strongly must we connect, as depicted in Figure 8.1, to achieve meaningful cooperation?

**Figure 8.1**   Networks of humans supported by communication networks.

And, what is even more difficult, to achieve cooperation between people with very different backgrounds?

There is a lot we can say about those questions, but let me first introduce some often-mentioned concepts or Networking Laws and see how they differ, so they are put in context and in perspective. I will avoid much of the mathematical symbols here, or put them double between brackets, since I know formulas can incite panic attacks in, for instance, sociologists and many other non-scientifically minded people.

First of all 'Network Effects' can be formally defined as: "How much does the *Value* change, proportional with the *Number* of people that participate in a network of connections/relations".

((Notation 'V'= *the Value*, and the *number* of reached people in the network is noted as 'N')).

These effects are the engines for New Power (connected demand side, see [7]) and they make networks grow and interconnect. Growth of Value can trigger growth in wealth, job creation and prosperity so it is worthwhile to study and implement them.

**I. Sarnoff's Law** *for broadcasting networks, radio, TV and media publication & distribution.*

The first networking effect is Sarnoff's Law, which states: the value of a one-to-many information distribution network grows proportionally with the number of readers (books, newspapers)/listeners (radio)/viewers (TV, movies). Or in other words: number of eyeballs of consumers, as shown in Figure 8.2. Such networks simply count viewers or views, without any regard for differences between or expectations of the users.

David Sarnoff has been able to convince with this 'law' the many local broadcasting stations in the US to interconnect and broadcast one or more nation-wide TV channels like ABC and NBC. More eyeballs, bigger audiences for the politicians and more value for the advertisers! See Figure 8.3. This network effect – more is better – is still the driving force behind 'publicity' and PR and is now in the process of being transposed onto online media on the Internet, where the media advertise themselves, insisting that us looking at their broadcasts is THE way of using the Internet.

I strongly disagree with this notion that spreading commercial content is the purpose of Internet, and I am getting more irritated every day by the bombardment of propaganda and commercials and their influence on the actual content of the Internet, swamping everything with sex-and-violence memes,

**Figure 8.2**   Families gathered around the TV set.

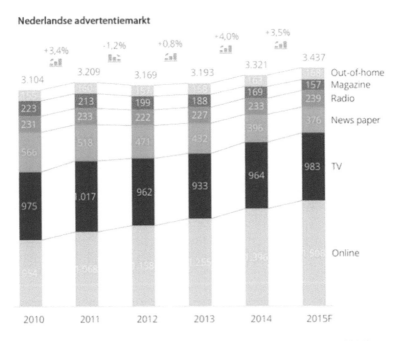

**Figure 8.3**   Advertising market of the Netherlands. Forecast online 2015: 1.5 billion euro's, double digit growth.

which tend to drag it down to the lowest common denominator level, like water. In contrast with what Internet did, it allowed its users, to find themselves the highest levels of their interests. Sarnoff's Law applies to one-direction messages with identical content, from a central source to as many passive, tranquilized 'information consumers' as possible – who, by the way, pay for the adds through taxes or the price of the products or services they are encouraged to buy and consume.

'Value' in this context means value for the broadcaster and commercial or public service advertisers. Politicians and business persons are fond of this form of network used to promote themselves or their products. They want to be in the centre of the public attention golden triangle of "screen, beer and bites" in which we live. And it is unfortunately also at the core of the business models of Google Search and Facebook etc. But as I've written above, users start to get annoyed by commercial breaks about things they do not want to buy, or lots of loud little, attention-grabbing and inescapable tiny films, monopolizing more and more space on our computer screens and phones, and for which we pay in subtle, indirect ways. This in my humble opinion (IMHO) is not a wrong intention of the advertisers, as long as their stuff is not forced unavoidably upon us, made obligatory so other choices are excluded. Business commercials and state propaganda have this same tendency to dominate.

((Sarnoff: Value is proportional to N viewers = $1+1+1+\ldots+1$; N times)). ((one-to-many, or 1:N))

Media Value **adds up**, which means the value counts up and is cumulative, irrespective of the diversity in demand or interests of the viewers. Of course, in some types of one-to-many publications, the number sold of a book or numbers of 'views' of a blog or 'likes' of a picture is a positive measure of income, popularity and/or appreciation, and can be used for ranking and attracting more viewers/readers. So Sarnoff's Law is a driver for growth of the World Wide Web too. Documents etc. are put online on the WWW to get more eyeballs on the page and reach and communicate with possibly interested people, who can still choose themselves.

## II. Metcalfe's Law for telecommunication networks with CONVERSA-TIONS and interconnection of networks.

The second network effect is Metcalfe's Law that states that the Value of a communication network growth is proportional with the square of the number of participants connected to it. Take for instance, a (mobile) telephone

**Figure 8.4**    People on the telephone.

conversation, See Figure 8.4, or an exchange of email messages. The reason is that each of the N people connected can talk to N (minus one, yourself) other people or computers.

((Value proportional to $N\hat{2}$ = NxN for N people/computers who can communicate with each other)).

Robert M. Metcalfe who defined this law, is also the inventor of the hugely successful (and disruptive) Ethernet Protocol. Developed to let computers share local area network resources, this is now also applied to sharing, using long-distance fiber optical cables. The value of such NxN telecommunication and computer-computer networks thus grows faster with N than 1:N media networks. It is also the reason why all fixed telephone and mobile telephone networks in the world are interconnected so that everybody can reach each other. It is also the driver for interconnection between datacom networks: the INTERnet and its attached services, which is, of course, a network of interconnected networks. It facilitates the SHARING of network infrastructure for digital transport. It drives peering and transit connections between Internet Service Providers (ISP's) and Carriers: a non-zero-sum game (win/win situation); a strong incentive to build and interconnect Fibre-to-the-X (FTTX) networks, so they can scale up and be hugely successful. The X denotes Home, Business, Premises or Farm.

Because of this Law the present total value of the telecommunication industry is multiple times that of the media industries combined in the world, contrary to what the media themselves would have us believe, including film and television celebs advertising themselves. To say it more compactly: "*Content is not king – communications is*" (quote of Odlyzko).

That people consider the possibility of connecting to billions of others as valuable can be concluded from the fact that they are willing to pay for calls and internet access. That growth of network Value is however, in practice, tempered by the fact that this law does not take into account the different strenghts of the links/relations between people. Or to put it more bluntly: "maybe I can talk to anybody in Brazil, but maybe I have nothing to discuss, not even to chat, except about the weather and to exchange some cat photos". The lower than square value growth has been derived in [8, 9] as the **Odlyzko-Tilly Law** which may be more realistic than Metcalfe's Law and Reed's Law but still a way bigger than Sarnoff's Law of media broadcasting.

$$((V \sim N * Log\ N))$$

### III. Reed's Law of GROUP MEMBERSHIP – *Value of Social (tribe internal) network*

The third network effect is in fact based on the "social networking behaviour" of people, as opposed to infrastructure investments and was formulated in 1999 [10] by prof. David P. Reed: of MIT. **Reed's Law** states that the value for users of big networks, and more specific: social networks grow in proportion exponentially with the groups of participants (members of tribes).

You can either be or not be a member of a closed (user) group/share some [14] or tribe, as depicted in Figure 8.5. So if there are N of those groups/clubs this gives 2 to the power N possibilities for memberships, growing exponentially when more groups are added.

Just try it yourself make a matrix of vertical N groups and make horizontal rows of persons who are (1) or are not (0) member of those groups. This charts the total number of possibilities and defines the total value of social networks that support tribes.

(Reed's Law: Value proportional to $2^N = 2*2*2*\ldots\ldots*2$, N times)

Why is being a member so valuable to people? It is the strong urge to belong, to be appreciated and be protected by fellow tribe members. That is, among

**Figure 8.5**  Groups.

other reasons, the drive to text/sms/Whatsapp/tweet all day and night with your friends: you confirm and are confirmed, socially, with group knowledge, you are member of the tribe, wherever you are at a given moment. It is very important for young people and they are willing to pay for it.

Reed's Law of Possibilities may have limitations, since not everybody can become a member of every club, because one's language, clothing and conduct must be acceptable to the other members of the clan. In addition, the fear of being refused or expelled can be significant. That is why members constantly confirm their mutual bonds. One does not need to be an anthropologist to see this. (Sometimes the recently initiated members even have to show themselves to be worthy of being a member by having to behave in a nasty way to non-members outside the tribe.) These social urges make the value growth of Reed's Law of group membership grow exponentially: faster and larger than either Sarnoff's or Metcalfe/Odlyzko-Tilly's.

There is something that has crept in almost surreptitiously onto the Internet, and which is implied by Reed's Law, that is, every human now can be member of more than one tribe/share some at the same time : "Multi Tribe Membership" (MTM). Before the Internet that was well-nigh impossible. Not so long ago if you were born in a village somewhere, in a certain family at a certain social level, or certain neighbourhood, the rest of your life could practically be charted until you died, no matter how hard you tried to raise your social status, for instance through education. Your own family members would frown if you started relations outside your social level, religion or tribe. And everywhere people would tag/judge you by your accent, dress or skill as "somebody from lower middle class in X".

You could go to the best schools, be a member of an elite sports club, listen to classical music and go frequent the opera and ballet; people would still remind you of your class/family/regional origins. The only escape was to move to the city or to another country, or join a different culture.

Now thanks to the Internet and other social media you can escape, be a banker in the morning, ride your Harley with your club in the afternoon and dance the tango at a samba school in another city. So you can be a member of multi tribes.

It is the exclusivity of these closed 'tribes' that attracts people. You join the group to be with people like yourself, with the same prejudices, dresses, who you can trust and feel safe with. The same cultural backgrounds and codes occur in Tweets, and by demonising outsiders. You can like this or not but the urge to be part of a group and be appreciated as a valuable member in that community is very strong, rational or not. And the urge not to be expelled is strong. In some parts of the world to be expelled from a village or clan can mean that you will not survive because access to water, food and shelter is taken from you. So Reed's Law of the value of memberships is strong, and may indeed be more valuable for society than the value of its communication networks.

### IV. Van Till's Law *for the value of COOPERATION & COLLABORATION*

The fourth network effect, even more strong and powerful than the first three as an engine for the Collaboration and Sharing Economy, as depicted in Figure 8.6, we now live in is **Van Till's Law** that states that the shared value for the Peer-to-Peer commons in which participants cooperate and collaborate in a network (which is more than just be a member) grows proportionally with N Factorial, [11–13].

(Van Till's Law of Shared Value:

V proportional to N! = N * N-1* N-2 * . . . . . . *2 *1).

This grows even faster than exponentially (viral) because this concerns the maximal (upper limit) number of combinations you can make by networked cooperation between N unique diverse and non-interchangeable) individuals, each contributing his/her first-class skills and crafts.

Example: with a deck of playing cards, each card unique, you can lay 52 ! different rows (combinations/sequences) of cards on the table: 52 ! = 52 * 51 * 50 * etc. With each card you choose from the deck that you lay down you have one less you can choose from. The 52 factorial combinations are a huge

**Figure 8.6**    Cooperation and collaboration.

number 3.1456688 E+69, which is bigger than the number of atoms estimated to be in our solar system. So this upper limit is impractical to estimate the value of cooperation. But it makes sense to show the richness of constructive possibilities when we connect and combine skilled people beyond the boundaries of their many closed tribes. The number of possible combinations of unique ideas is limitless, only constrained by lack of imagination and cultural and conservative prejudices and silo's with vested interests. These are the boundaries we have to cross if we want to create value and give our children a future.

Crucial point of Van Till's Law is not the upper bound of its Value, but that cooperation and co-creation by participants, with the best [16] contributions and their first-class skills, whatever their background, online and off-line, works in a ***multiplicative*** way to create value, as opposed to ***adding up***, as with media broadcasting.

This supports the popular slogan of the P2P Open Share Economy: "To share in a smart way is to multiply value for all contributors" (in Dutch language: "Delen is vermenigvuldigen". This is a pun in Dutch, based on the double-meaning of 'delen', as both 'sharing' and 'division', the latter being the opposite of 'multiplication'; it seems to say 'division is multiplication', which is a nice paradox).

If a valuable person joins your team, who can really DO something extremely well and this contribute as promised, it can multiply the value of the whole team. Besides, the group can act as a magnet to others, thus speeding up and progressively enhancing the learning process.

P2P 'Sharing' implies in this context not only physical resources and tools but also of practical knowledge and information about solving problems. Knowledge and information does not diminish if you share it and is therefore abundant, non-scarce and thus inexhaustible, see Figure 8.7.

The business community is noticing the same four network effects as phases in Social Media binding of clients. For instance Gaurav Mishra of 20:20 WebTech identified four "phases" in connecting with clients and groups of clients: 1. Content 2. Collaboration 3. Community 4. Collective Intelligence, see Figure 8.8. With each step the effort becomes more invisible, but also more difficult to weave. And yes: curators, conversations, co-creation and recommendation/reputation/trust play critical roles.

## 8.4 The Telescope Metaphore

So from the Network Effects we may conclude that successful cities not only are built on *scale* (number of inhabitants) but also on *connectivity* & *scope*: depth of skills of top specialists which are co-creating by way of networking together and by cooperation.

**Figure 8.7**  Many small fish can together form something big.

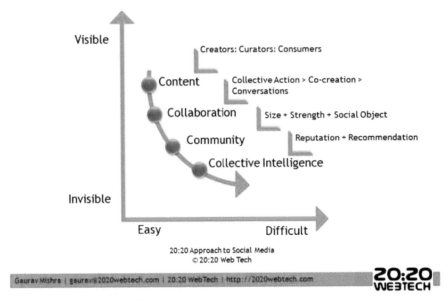

**Figure 8.8** Approach to social media.

*Source*: Gaurav Mishra, http://202webtech.com

These skills must be diverse and unique, like the DIFFERENT playing cards that are combined to make a strong hand. What is the reason for this need for diversity in cities? An explanation can be found in the "Telescope Metaphor" which I did define in 1998 [18]. Centuries ago people started to make telescopes to better look at the stars. Until quite recently it was assumed that a bigger telescope with larger lenses would 'catch more photons' from a distant star and would have a better resolution. Radio-wavelength astronomers found out that this is not the case. By linking a number of small radio dishes, each looking at a star, and processing the combination of signals, they get a much better image than from one large telescope. The resolution of the whole *Network* of small telescopes does depend on the distance between the two little telescopes which are the most far apart! As a result astronomers now 'network' their telescopes in arrays for combined observation from far apart locations on earth or even by linking with one telescope on a satellite.

Telescopes where invented and improved in Holland and Venice, who both employed them to have early info about cargo ships that returned from Asia, so they could buy stock from that ship just before it arrived. Later ever bigger telescopes where used to look at the stars and planets. Bigger until the glass lenses and mirrors nearly started to collapse. Size of most things can reach an upper bound. That applies to buildings, companies and cities too.

The solution is to split up the large telescope into a number of inter-connected (networked) small ones that effectively function as **one** telescope. This setup is for instance employed by the huge LOFAR array radio tele-scope, which in the Netherlands and Germany employs 20,000 interconnected antennas.

The same engineering trick was done with mobile (phone) networks. Instead of installing ever more powerful and more frequency carrying antennas they reversed the direction to install many small cellular antennas with short ranges Nordic Mobile Telecoms (NMT) and non-overlapping frequen-cies, from which the base stations where interconnected with fixed leased lines. Such cellular networks can handle many more mobile subscribers and heavy data-transport traffic. Later generations of communication networks for mobile users have the same cellular architecture of small connected cells.

Coming back on radio telescopes and to be more specific, the **Resolution** of the synthetic telescope, we know that it is related to the distance between the most widely apart little telescopes (see Figure 8.9) and the **Sensitivity** (number of bits/dynamic range of the signals) which is related to the Number of the small telescopes. The effectiveness of the total setup is depending on

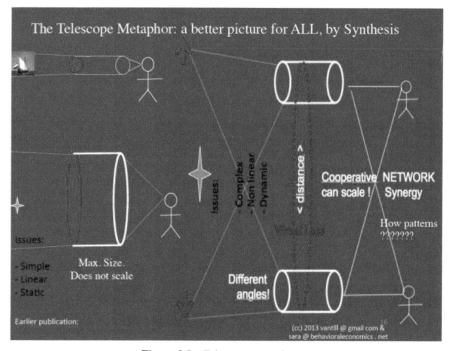

**Figure 8.9**　Telescope metaphor-1.

what processing and networking power is behind the telescopes. We see not with our eyes but with the lens of our brain!!

So this 'telescope metaphor' also applies to our body. We see depth by combining the *different* points of view of two eyes. Cover one eye to see what is lost by sensor ship. Our view is improved if we network with others who are far away from us, and who can look at the same scene from a *different angle or perspective*. Each participant gains in this process. All images which are made (processed and presented inside different brains) gain resolution.

This is the central idea of the new 'Networked Democracy': we all gain by communication and discussion from most different views, even the most far-fetched or un-welcome ones.

Now we can ALL see things much clearer: networked together, individual & connected! So there is an incentive for all contributors to network and collaborate together, each gets an improved overview in return for their own contribution. Personal interests are connected to the shared interests and fed back to the personal ones, in a virtuous circle. This fire the chimney effect mentioned above.

If such skilled persons and constructive teams are woven together, brought together by Networkers and Btwieners [15] and a flock or maze builds up, it suddenly is possible to have hundreds of thousands of angry women demonstrating with pink hats on squares all over the planet, who behave and act together like one non-violent coordinated stable living organism.

It turns out that Btwieners, with the help of smartphone networks, can bring stability by binding people from multiple closed groups together across boundaries, instead of dividing them by hate and conflicts erupting from multi-ethnic tribes clashing with each other.

As stated above, besides gains in resolution of a group of very different positioned people and higher perceptive sensitivity for all by having a large number of people interacting and cooperating we also need a particular kind of scalable network structure to correlate, process and distribute the information.

My prognosis is that the structure of the interconnections between people that will have the emergent property of correlation, distributed memory of patterns and "collective intelligence", will be similar to Weavelets, defined in [19] and [20].

You should not be surprised that this Weavelet pattern will turn up in Brain Research, Social Media core structures and NSA -like psy-ops big data systems.

To explain this proposed structure, please notice that in order to synthesise a picture from a wide combination of sources, as shown in Figure 8.10, all

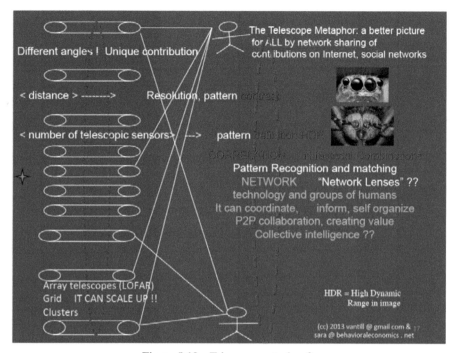

**Figure 8.10** Telescope metaphor-2.

the different information streams must be connected to each other. This is done in the brain of insects with all the facet eyes, linked together. And it is done inside the head of spiders who have a number of separate eyes on their head. One of the most efficient ways to connect every sensor stream is to do that in a number of layers. If certain weighting factors for amplitude and phase are used, compare that to Twitter likes with comments, after a few layers we have transformed the spatial image into a transformed one, which can be used to correlate and recognize previous knowledge. Such a construction is fast, parallel, and resilient against damages and noise, it can store knowledge in the form of holograms which replicate copies all over the network. My conjecture is that such networks construct "synthetic apertures" and can be compared to a "lens" which can resolve knowledge and lets people learn together, by mixing view into a larger whole in which they can participate and benefit. This way we can build cities which have a collective intelligent community.

In reality human brains, social networks, 'deep thought' AI systems, search engine databases, WikiPedia, Facebook and their Chinese counterparts have 'neurons' with very many input and output connections instead of 2 in and 2

"Weavelet" Network Structure for P2P Collaboration between people (= nodes)

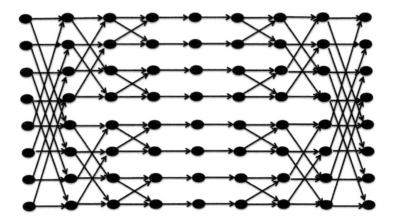

Derived from Cooley-Tukey algoritm for fast orthogonal transforms by Jaap van Till, 2013
Such connectivity between people can lead to "Collective Intelligence with Distributed Authority"

**Figure 8.11**   Fractal scaling weavelet networks that form a parallel LENS-LIKE structure, capable of correlation, pattern recognition and distributed memory.

out of the depicted Weavelet of Figure 8.11. People who are active in social networks usually have several hundred other people they frequently talk to and up to a hundred with whom they actively cooperate with. I introduced the Weavelets here to emphasize the role and importance of 'connections' for science, social cohesion and consensus.

The value and wealth generated by the interconnection between real top craftsmen and -women is too powerful to resist. This is also my recipe for smart cities and chains of cities beyond that:

> *Interconnect across boundaries the members of selfish and proud sects and tribes together into a Network, by way of problem solving, by which you need each other's specialists, and by constructive shared goals that are impressive. Cathedrals where built in that way, by craftsmen who respected each other's work. We can do that in the P2P Commons of the Collaborative Networked Economy too, but we first must learn to form well working Weavelets, constructive groups with "collective intelligence".*

I hope that the Fourth Network Effect of collaboration will get us out of the recent recessions at last, and will make Cities and Chains of cities bloom. May the Fourth Law *Force* of Synergy & Synergetics be with us !!

## 8.5 From Mega-Metropoli to Chains of City Areas, Example: The Eur-Asean Trade Route Called "Corridoria"

A large part of the world population has been attracted by the big cities to find work and have health services etc which are available there and not in the villages where they were born. This has resulted in mega-metropoli which have difficulty to cope with food, sewage and air pollution which reaches the boundaries of size when many extra millions of people keep coming every year to each city and the rural agricultural population diminishes. And unfortunately smart cities in most countries think themselves as being in the centre of the universe to which other smaller cities in the area should connect in a star shape. For instance Antwerp, Rotterdam and Amsterdam each think in such a self-centred way, while they are IMHO part of a city called EuroDelta.

In my opinion it is inevitable that just like with mobile phone networks and telescopes a policy & engineering reversal will be implemented to solve this problem. That means a chain of city area's (and mega-poli) connected in a string. In this way the condition of resolution [different & far away contributors can join, like in the travels of Marco Polo, bringing silk and emeralds to Venice and glass to the Khan] is fulfilled, and condition of sensitivity is also fulfilled (number of inhabitants in the whole chain) and can grow further.

This was the first drawing (Figure 8.12) I made of Corridoria on August 14, 2013 and published on my blog. The trajectory roughly followed the beds of both River Rhine and Danube which make sense since for transport you thus can ship things without great heights to climb or slide down. Armies preferred such routes as well as pilgrims, train tracks and couriers with messages on horseback. This route through Europe is now booming with growing traffic of lorries. Later that year I added the route from Istanbul to Shanghai (Figure 8.13), in China also along a river, which improves the just mentioned conditions even further. There are a lot of other reasons for this trajectory, which is detailed in [21]. Recently the Chinese Government has unfolded a huge plan of economic cooperation westward in several new Silk Route routes, mainly with train transport from China to Rotterdam, and other cities in the EU.

But these developments are more than only train tracks. There are, maybe in response to the turbulence on the UK and USA side, a number of economic pacts in Eur-Asia unfolding: *CTSS: De Turkish Council* and *OBOR: One Belt, One Road* started from the Chinese side as mentioned.

What would make more sense for commerce and prosperity than a link between EU – CTSS – OBOR? This is in essence what Corridoria can start to be!!!

**Figure 8.12**   First sketch of "Corridoria", string of cities, 2013 and derived from noticeable growth of traffic and prosperity.

I would like to add two things here about Corridoria for the sake of the argument of this chapter that connected people do more than simply discuss: (1) They cooperate (Van Till's Law which creates value) and they form network structures that exhibit emergent "collective intelligence" (condition as said before) to harness prosperity and innovate. First of all the big cities on the Corridoria line will prosper and bloom with their inner city infrastructures data centres and services now connected to all inhabitants on the chain. By traveling and university exchanges scholars and students will meet others on this Eur-Asian chain city and learn together.

Something similar has happened along the silk routes around the years 1200 in Middle Asia. Scholars followed the example of the very tolerant court of Persian Emperor Darius the Great (ca 650 BC) who understood that a MIXTURE of very wise scholars would provide knowledge and education to be successful. These scholars from China, India and European origin around 1200 met in cities like Merv and Samarkand and established "schools" from

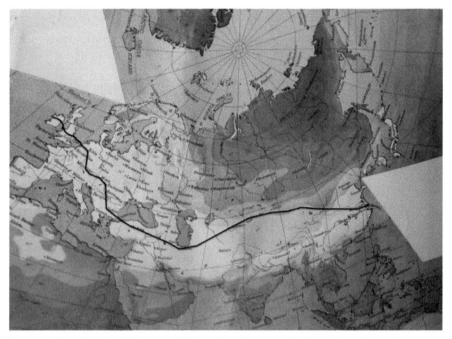

**Figure 8.13**  "Corridoria" String of Cities, from Galway – Dublin – Euro Delta – Istanbul – Iran – Shanghai. 2013. From the Chinese side they are developing rail transport towards Europe (OBOR Strategy).

which a very large part of the present culture and knowledge in the world originated [22]. I hope we can repeat such huge jump in Enlightenment by letting present scholars and students meet in cities along Corridoria.

(2) A second point is that "Corridoria" is not structured simply as a line. Connected to the backbone are a multitude of smaller cities and towns in a resilient network structure as depicted on my following photo. I call that structure "Gaia's Necklace" which will span all of Europe and Asia. That network structure will provide the Global Brain with a weave of weavelets, thus fulfilling the condition mentioned.

### Notes and Remarks
ICT networks do indeed provide very powerful support for cooperating people. We should however keep in mind that there is no substitute for human contact, face-to-face in real life.

I predict however that smart networking using internet and telecommunications will improve the images we can gather about reality by:

**Figure 8.14**   Gaia's Necklace of city connectivity towards the global brain.

A. *Contrast in perception* for all: related to the distances of the networks
B. *Sensitivity of images* and perceived patterns: related to the number of inhabitants who are connected and can freely move information.
C. *Level of consciousness* of society: related to the density of weavelet connectivity.

## 8.6 Conclusion

As the recent article [23] presents, we as human society are faced with an urgent problem: Dynamic Complexity resulting from (tele) interconnecting everything – with – everything, which makes controlling & dictating countries from a central point impossible without simplification and fragmentation of groups.

Commercial interests have developed tools on Internet to profile people and by confirmation of their views, filtering of the information they get and nudging their beliefs in an intended direction create "bubble" belief systems of target groups. The consequence of this is isolation, hate and conflicts between such fragmented "tribes", that are no longer able to understand and communicate with each other, even within cities.

A possible direction of solution of this problem is to re-connect people within cities and between cities, using tools to synthesise their views and solution skills.

This can be done with 'synthetic aperture' tools for connected people as applied to modern array telescopes. Such synthetic aperture can be constructed by a structure I have called "weavelets", which are fast, resilient, can scale up and can yield information which is replicated over all participants, like a hologram. The incentive to join such connected communities is in the valuable synergy which is created together in which all participants can share, cooperate and use. This is a possible way forward in which we can build world-wide communities that live insides and outsides chains of cities.

## References

[1] Khanna, P. (2016). *Connectography: Mapping the Future of Global Civilisation*, Available at: http://www.paragkhanna.com/connectography/

[2] Khanna, P. (2017). *Global Gateway on* Dubai, *Video Features Commentary on the City's Resilience and Economic Strategy, CNNi Video*. Available at: https://app.frame.io/f/VEm0ptIj

[3] Rifkin, J. (2014). *The Zero Marginal Cost Society – the Internet of Things, The Collaborative Commons and the Eclipse of Capitalism*. New York, NY: St. Martin's Press.

[4] Van Till, J. (2014). *Lecture Slides about Social Networking*. Available at: https://theconnectivist.wordpress.com/2014/11/08/slides-of-my-social-networking-masterclass-workshop-at-osdc2014/

[5] Csermely, P. (2006). *Weak Links, Stabilizers of Complex Systems from Proteins to Social Networks*. Berlin: Springer.

[6] Van Till, J. (2015). *Blog about the Four Network Effects*. Available at: https://theconnectivist.wordpress.com/2015/03/25/np9-engines-for-the-new-power-the-four-network-effects/

[7] Van Till, J. (2014). *Blog about New Power*. Available at: https://theconnectivist.wordpress.com/2014/12/20/new-power-on-the-rise-user-side-synthecracy/

[8] Odlyzko, A. M., and Benjamin, T. (2005). *A Refutation of Metcalfe's Law and A Better Estimate for the Value of Networks and Network Interconnections*. Available at: http://www.dtc.umn.edu/~odlyzko/doc/metcalfe.pdf

[8a] Briscoe, B., Odlyzko, A., and Tilly, B. (2006). *Metcalfe's Law is Wrong*. Available at: http://spectrum.ieee.org/computing/networks/metcalfes-law-is-wrong

[9] Reed, D. P. (1999). *"Weapon of Math Destruction: A Simple Formula Explains Why the Internet is wreaking Havoc on Business Models," Context Magazine, Spring 1999*. Available at: http://web.archive.org/web/

20080526050751/http://www.contextmag.com/setFrameRedirect.asp?
src=/archives/199903/digitalstrategy.asp

[10] Van Till, J. (2004). *"The Fourth Internetworking Law"*. Available at:
https://tnc2004.terena.org/core_getfileeee3.pdf?file_id=389

[11] Van Till, J. (*2008*). (In the Dutch language) *Chapter in SURF WTR Book 2008 and Published in Netkwesties*. Available at:
http://www.netkwesties.nl/788/engines-for-the-new-power-the-four.htm

[12] Schimmelpennink, L. van Till, J. Kuitert, K. Lindler, L. Sala, L. (2015). (In the Dutch Language) *"De Verbonden Stad" – Essays over de Stedelijke Omgeving in de Context van Diversiteit en Mobiliteit*. Available at:
http://www.boekcoop.nl

[13] Van Till, J. (2014). *Definition of a Sharesome*. Available at: https://theconnectivist.wordpress.com/2014/12/28/definition-of-a-sharesome-nl-een-deelzaam/

[14] Van Till, J. (2014). *Definition and Importance of Btwieners*. Available at: https://theconnectivist.wordpress.com/2014/11/08/slides-of-my-social-networking-masterclass-workshop-at-osdc2014/

[15] Van Till, J. (2014). *Van Till's Principle (VTP) of Well Connectedness*. Available at: https://theconnectivist.wordpress.com/2014/07/14/van-tills-principle-of-maximum-well-connectedness/

[16] van der Lee, R., Taffijn, P., and Anderen, E. V. (2015). (In the Dutch Language) *Het Kantelings Alfabet – Verandering begint met Delen –; Een co-Creatie van een Divers Gezelschap van Kantelaars, Dwarsdenkers, Verbinders, Friskijkers en Verkenners E*. Available at: info@dealfabetboeken.nl

[17] Van Till, J. (2016). *The Eyes of the Earth and The Telescope Metaphor*. Available at: http://www.vantill.dds.nl/democracy.html

[18] Van Till, J. (2013). *The Telescope Metaphor Sheets*. Available at: https://theconnectivist.wordpress.com/2013/04/28/p2p-connectivism/

[19] Van Till, J. (2013). *Introduction to Weavelets and the Weave*. Available at: https://theconnectivist.wordpress.com/2013/05/12/power-shifts/

[20] Van Till, J. (2013). *More General Vision on Future Networking with Examples from Biology*. Available at: https://theconnectivist.files.wordpress.com/2013/01/sheets-weavelet-jvtdef6.pdf, 2013.

[21] Van Till, J. (2014). *The List of Cities on the Corridoria Chain of Cities*. Available at: https://theconnectivist.wordpress.com/2014/04/13/the-list-of-booming-city-regions-on-the-corridoria-trail-april13-14/

[22] Starr, S. F. (2013). *Lost Enlightenment – Central Asia's Golden Age, from the Arab Conquest to Tamerlane.* Princeton, NJ: Princeton University Press.

[23] Helbing, D., Frey, B. S., Gigerenzer, G., Hafen, E., Hagner, M., Hofstetter, Y., et al. (2017). "Digitale demokratie statt datendiktatur," in *Unsere digitale Zukunft,* ed. C. Könneker (Berlin: Springer), 3–21.

[24] Ahmed, N., and Rao, K. R. (1975). *Orthogonal Transforms for Digital Signal Processing.* Berlin: Springer.

## Biography

**Jaap van Till**, retired, was professor at the HAN University of Applied Sciences in Arnhem (NL) and parttime professor for Corporate Networks at the Delft University (NL) and the Tech. University of Kaunas (Lithuania). He frequently taught at post-graduate courses and business schools, like the Institut Theseus in Nice (France), at Amsterdam University, Kumasi Technical University (Ghana) and University of Indonesia.

Prof. ir. Jaap W.J. baron van Till (73) is a network engineer with a degree in signal processing & pattern recognition at Delft University. He has experience in the Telecommunication, Computer Network and Internet-infrastructure fields, and is working at the cutting edge of new disruptive innovations, social networks and emerging P2P community tools, Fiber-to-the-Home and ICT technology policies and regulatory government policy. He learned from his projects as a "Network Architect" in the corporate, laboratory and factory networks of Akzo Nobel all over Europe. Later he helped to design large corporate computer networks for businesses, government ministries, the NL NREN (national research and education network) SURFnet and the Netherlight lambda network. Jaap was member of the Technology and Science Council (WTR) of the SURF Foundation which oversees the knowledge infrastructure for 1.5 million smart people in the Netherlands.

His present research interests are: digital infrastructures, P2P value co-creating commons, Collective intelligence with distributed authority, and how to cope with complexity and the unexpected. Jaap does consulting at Tildro Research, Rhenen, NL and is non-executive board member of the NDIX, the Netherlands- German distributed Internet Exchange network.

# 9

# Smart Cities – A Panacea for the Ills of Urbanization: An Indian Perspective

M. D. Lele

Vishwa Niketan Institute of Management, Entrepreneurship, Engineering & Technology, Mumbai, India

## Abstract

Cities across the globe occupy 3% of the earth's land surface, house half of the human population, use 75% of the resources and account for 2/3rd of all energy usage and greenhouse gas emissions. Reliable estimates have pegged the urban population of the world by 2050 at 70%. To deal with the problems of urbanization such as population explosion, high level of pollution, traffic chaos, huge scarcity of houses, inadequate water supply, heaps of garbage, poor infrastructure, security concerns, unemployment, overpowering political set up, etc. the concept of inclusive Smart cities, evolved in the European Union and has now spread across the continents. Thus a new approach to take on the colossal challenges of this century has emerged.

It is expected that the Smart Cities will address public issues, employing Information and Communication Technology (ICT) based solutions, on the basis of municipality driven multi-stakeholder partnership. The need to strengthen or extend citizens' participation in initiatives by Municipalities and emphasize establishment of more refined and accessible governance structure has been realized. While the concept of Smart Cities is still under development and evolution, it has caught on across the globe and has spearheaded a new approach to look at urban development and management. Cities have their own requirements and peculiar limitations necessitating public centric strategies, appropriate and economically viable technology and funding mechanisms.

Cities accommodate nearly 31% of India's current population and contribute 63% of GDP. Urban areas are expected to house 40% of India's

*Breakthroughs in Smart City Implementation,* 215–242.

population and contribute 75% of India's GDP by 2030. Indian Government would earmark US D 1054 million to fund the development of a hundred Smart Cities with a view to address various problems arising out of an accentuated rate of population influx. The Mission will have duration of five years (FY2015-16 to FY2019-20). The city level Pre-takeoff Plans in respect of quite a few cities are stimulating investment by engaging identified stakeholders i.e. ICT businesses, real estate developers, infrastructure providers, retail business and transport operators to support Public Private Partnership (PPP) projects and actions. The objective of Indian Smart City initiatives is to increase the resilience of the city. It is proposed to develop the 100 Smart Cities as Satellite Towns of larger cities by modernizing existing mid-sized cities. The aim is to promote cities that provide core infrastructure and give a decent quality of life to its citizens, a clean and sustainable environment and application of 'Smart' Solutions. The focus is on sustainable and inclusive development and the idea is to look at compact areas, create a replicable model which will act like a light house to other aspiring cities.

The new initiative of the Government of India to make cities capable of facing the 21st century challenges with the generous reliance on the latest frontiers of technologies available, serious efforts towards capacity building of the personnel manning the projects and above all employing methods to make them financially sustainable with little aid will go a long way in creating a Resurgent Urban India. The importance of deploying advanced technologies for solving the problems of city governance with ease cannot be undermined. But at the same time while embracing newer and smarter technologies to offer ease of doing business, care should be taken to make available services to the lowest denominator of the society too, in a way he/she is comfortable with.

**Keywords:** CIDCO, BIS, PPP, EPC, JNNURM, GDP, ULB, CSS, SCP, GIFT city, NGO, SPV, BRTS, NOC, SCADA, NCT, Smart City, Lavasa, Navi Mumbai.

## 9.1 Introduction

### 9.1.1 Background

Cities across the globe occupy 3% of the earth's land surface, house half of the human population, use 75% of the resources and account for two-third of all the energy usage and greenhouse gas emissions. Reliable estimates have pegged the urban population of the world by 2050 at 70% [1–3].

Moreover, the twenty-first century is coming to be known as the century of cities. Further, the UN forecasts Asian cities will grow faster than the western cities in the next few decades [4].

Plagued as they are by population explosion, high level of pollution, traffic chaos, huge scarcity of houses, inadequate water supply, heaps of garbage, poor infrastructure, security concerns, unemployment, overpowering political set up, etc.; they will have to find methods to deal with the problems, have better energy, water and waste management and cope up with the rising urban population and high densities through innovative planning and policies. To deal with the problems of urbanization the concept of inclusive Smart Cities, evolved in the European Union has now spread across the continents. Thus a new approach to take on the colossal challenges of this century has emerged.

## 9.1.2 Evolution of the Smart City Concept

The origin of the Smart City Concept lies among the widespread ills seen due to rapid urbanization. The impact of all this growth on space, environment and quality of life will be, to say the least, tremendous. The provision of infrastructural facilities required to support such large concentration of population is lagging far behind the pace of urbanization. To deal with the problem it was felt by the European Union that a fresh look needs to be taken and a new approach to be thought over. It believed that investments should be made in human, social capital, traditional transport and modern (ICT) communication infrastructure fuel sustainable economic development and a high quality of life should be provided by engaging wise management of natural resources, through participatory action [5].

It is expected that the Smart Cities will address public issues, employing Information and Communication Technology (ICT) based solutions, on the basis of municipality driven multi-stakeholder partnership. The need to strengthen or extend citizens' participation in initiatives by Municipalities (democratic organizations at primary level) and emphasize establishment of more refined and accessible governance structure was realized. British Government customized the EU Smart City concept to make it more focused to develop core competencies and achieve part of its millennium goals. These goals were [6]:

Goal 1: Eradicate extreme poverty and hunger.
Goal 2: Achieve universal primary education.
Goal 3: Promote gender equality and empower women.
Goal 4: Reduce child mortality.

Goal 5: Improve Maternal Health.

Goal 6: Combat HIV/AIDS, malaria and other diseases.

Goal 7: Ensure environmental sustainability.

Goal 8: Develop a global partnership for development.

The Department of Business Innovation and Skill (BIS), London wanted the Smart City to be an integrator of Physical, Digital and Human Systems in a built environment to deliver sustainable, prosperous and inclusive future for its citizens. Thus began the evolution of this idea into concrete action [7].

EU's initial focus was integration of Citizens and Governance using ICT in terms of effective knowledge sharing with stake holders, creating common framework to develop citizen insight, linking planning design and operation of systems, keeping the citizens' needs in mind. As most of the EU member states had some form of extended participation of its citizens in municipal governance, the focus shifted towards ICT products driven by different vendors and vendor driven protocol. During this time products, protocols and partnerships in public and private sector evolved.

While the concept of Smart Cities is still under development and evolution, it has caught on across the globe and has spearheaded a new approach to look at urban development and management with underlying principles such as attracting young wealth creators, engaging in constant physical renewal, creating a unique and strong city identity, being connected to other cities, inculcating innovation and having strong political and administrative leaders. Finally, based on the above analysis, the definition for a smart sustainable city as approved by the ITU-T FG-SSC is as follows:

"A smart sustainable city is an innovative city that uses information and communication technologies (ICTs) and other means to improve quality of life, efficiency of urban operation and services, and competitiveness, while ensuring that it meets the needs of present and future generations with respect to economic, social and environmental aspects".

### 9.1.3 Labeling a City 'Smart'

A Smart City was seen by EU as a city seeking to address public issues via ICT based solutions on the basis of a municipality based multi-stakeholder partnerships. Though the emphasis/focus of Smart Cities in the foreign context has been on technology (IT and communication technology) it is not the singular factor that contributes to making a city smart. For cities are complex and multidimensional. Urbanization influences all aspects of society and economy and its growing connections with technological revolution make socio-economic processes complex and interesting. Cities have their own requirements and peculiar limitations necessitating public centric strategies, appropriate and economically viable technology and funding mechanisms.

## 9.1.4 Global Appeal

The new concept of doing the same things, but differently and efficiently has been received quite well across the geographies. The experiences shared here offer good learning lessons for replication wherever possible with a customized approach suiting the local conditions [8].

### 9.1.4.1 Rio-De Janeiro

The idea of having a Single nodal centre was implemented with success. PPP model was adopted. Centre's representatives were appointed to operate 30 centers. Lack of infrastructure was overcome by re-building old projects.

Administrative and environmental challenges were solved by cutting energy consumption by 20%. This was possible due to installation of an Energy Management System under the complete control of the Central Command Centre (COR) (Figure 9.1). The local government has been working on initiatives on energy efficiency, with the challenge of identifying impact solutions for the city of Rio. The Program for Modernization of the Street Lighting Network, for example, includes geo-referenced mapping of the street lighting network of the city, cataloging points of light and equipment. The

**Figure 9.1** Central Command at Rio [9].

program also provides for the development of a Master Plan for Street Lighting and the replacement of technology with LED and solar bulbs [9].

Contractor paid for additional construction rights to exploit commercial space. Thus the Olympic Park was developed.

A series of structural interventions in the transport and mobility area is in progress in Rio de Janeiro. The technology projects that are in progress include modernization of the infrastructure of the traffic equipment network such as traffic lights, variable message signs, cameras, and other sensors. Other initiatives are also being deployed, such as "Digital Traffic," which monitors and reports in real time the path of the main routes and alternative routes in the city. Commuters now receive traffic alerts and re-routed directions through social media before congestion becomes problematic.

### 9.1.4.2 Amsterdam

The 5 themes that were set by the organizations were Living, Working, Mobility, Public Facilities and Open Data. 'Living' involved Citizens' $CO_2$ awareness and dwelling refurbishment. The 'Working' theme revolved around shared co-working spaces. Mobility embodied sustainable means of transport. Provision of public facilities was through strategic role assignment to the private player. Data encryption was done to fuel the information society. In the case of Amsterdam (Figure 9.2) three types of data of roughly 4,00,000 buildings formed the basis of the Decision Support Environment:

1. Building specific information (age, surface, energy label, type, function etc.)
2. Energy consumption (electricity, gas, and heat consumption)
3. Energy potential (solar and wind potential as well as geo-therma land thermal storage capacities) The Decision Support Environment enables cities to interact with stakeholders and to discuss future energy scenarios based on factual data: a fundamental basis to develop the right energy transition strategy and involve all stakeholders and decision makers [10].

### 9.1.4.3 Seoul

Smart Seoul 2015 was adopted to overcome the limitations of u-Seoul which applied ICTs only to existing 'traditional' city infrastructure. U-Seoul improved the delivery of services such as transportation and safety, but failed to produce material improvements in the quality of life enjoyed by Seoul's citizens. Smart Seoul 2015 is a more people-oriented or human-centric project; and Seoul now aims to implement not only as many smart technologies as

**Figure 9.2** View of Amsterdam [10].

possible, but also to create a more collaborative relationship between the city and its citizens. 'Smart Devices for All' were made for citizens to communicate with government bodies and other administrative offices. "Smart work centre" & "Smart Metering project" were made to manage work from the common place and reduce the city energy consumption by 10% [11].

### 9.1.4.4 Masdar

The city has the traditional Arabic design having an urban realm. (Figure 9.3) It has optimally-oriented and well-integrated transport systems catering to the low rise high density living. The pedestrian friendly town providing high quality living has shaded walkways and a network connecting strategically located parking lots with driverless point to point personal rapid transit system. It is now exploring metro lines, light rail and electric bus operation options.

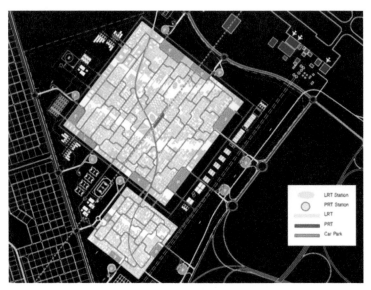

**Figure 9.3**   Concept plan of Masdar [38].

### 9.1.4.5 Songdo

Central park canal uses seawater instead of fresh water whereas reused grey water reduces potable water demand by 90%. Further while plumbing fixtures target a 20–40% reduction the storm water runoff will be reused. (Figure 9.4) Vegetated green roofs will reduce the urban heat island effect.

**Figure 9.4**   City scape of Songdo [38].

### 9.1.4.6 Singapore

Traffic data is aggregated, integrated and disseminated at the ITS Centre for traffic monitoring and management. The data is also collected for traffic analysis and planning. Real time and localized traffic information is disseminated through websites, radio broadcasts and smart phones. To reach out to the wider public and leverage on the private sector's innovations as well as their broad distribution channels, traffic information is also disseminated through the private sector's products and services. Singaporeans receive traffic predictions with an accuracy of 90 per cent [12].

### 9.1.4.7 Dubai

Dubai is to use flexible energy from renewable sources. Dubai has announced plans to build the world's largest concentrated solar power (CSP) facility by 2030 as part of a wider push towards renewable energy. The plant, located within the Mohammed bin Rashid Al Maktoum Solar Park, is set to generate 1,000 megawatts of power by 2020 and up to five times that amount by 2030. The new CSP plant is part of the emirate's plan to shift to renewable – Dubai is hoping to generate 7% of its total power from clean energy by 2020, followed by 25% in 2030, and 75% by 2050. Rooftop solar schemes could provide another 2500 MW which is 20 percent of domestic power. It was proposed that by 2014, all buildings in Dubai must be energy efficient according to new green building codes.

As of 22/01/2017, all qualified bidders were requested to present proposals for a 200 MW concentrated solar power, CSP, plant, the fourth phase of the Mohammed bin Rashid Al Maktoum Solar Park in Dubai, Construction of the 800 megawatt phase 3 of the Mohammed bin Rashid Al Maktoum Solar Park in Dubai was to start at the end of January 2017 following the award of the engineering, procurement and construction, EPC, contract for the project [13, 14].

## 9.2 Scenario in India

Cities accommodate nearly 31% of India's current population and contribute 63% of GDP. Urban areas are expected to house 40% of India's population and contribute 75% of India's GDP by 2030 [15, 16].

Five states in India are projected to urbanize beyond 50% by the next two decades. The pace and spread of urbanization is not uniform. Maharashtra with an urban population of 42 per cent (41 million), and the least urbanized state, Assam with 13 per cent in 2001 indicate this inter-regional variation [17].

India is facing an unprecedented scale of urbanization with 700 million people likely to move to cities by 2050 out of an estimated 1900 million people [18]. Of these 576 million people will live in 68 metro cities pointing to the fact that the urban system is skewed towards bigger cities [19].

Infrastructural requirements will grow exponentially in the years to come in this fourth largest world economy by GDP [20], having second highest GDP amongst developing countries. Magnitude of the problems of housing shortage, deficient water supply, insanitary conditions, solid waste management, traffic congestion and environmental degradation is humongous. Moreover, phenomena like haphazard peri-urban development, ribbon development and multiplicity of agencies and legislations, and lengthy legal procedures to mitigate unauthorized developments pose severe challenge. To cope up with this nightmare, it was believed that we need to instill 'smartness' in at least a hundred cities.

The Govt. of India in its draft report [21] states that Smartness in a city means different things to different people. It could be smart design, smart utilities, smart housing, smart mobility, and smart technology, etc [22].

The British Standards Institute's definition goes thus [23]: "Smart Cities" is a term denoting the effective integration of physical, digital and human systems in the built environment to deliver a sustainable, prosperous and inclusive future for its citizens." In other words, a Smart City is one which uses advances in technology for efficient working of the city and essentially embodies comprehensive planning, inclusive policies, efficient transport, sustainable economy, environment conservation and above all, proactive governance [24].

### 9.2.1 Government of India's Action Plan for Smart Cities

As a sequel to the considerably successful JNNURM project [25] aimed at transforming metro and Class I cities, Prime Minister, Mr. Narendra Modi announced in his maiden national budget in 2014–15 that Indian Government would earmark US D1054 million to fund the development of a hundred Smart Cities [26]. The Mission will cover 100 cities and its duration will be five years (FY2015-16 to FY2019-20). The Mission may be continued thereafter in the light of an evaluation to be done by the Ministry of Urban Development (MoUD) and incorporating the learning into the Mission. The Smart City Mission will be operated as a centrally sponsored scheme (CSS) and the central government proposes to give financial support to the mission to the extent of US D7165 million over five years i.e. on an average US D149 million per city per year. An equal amount, on a matching basis, will have to be contributed by the State/ULB; therefore, nearly US D14923 million of Government/ULB funds will be available for Smart Cities' development.

After the Stage 1 of the challenge, each potential Smart City will be given an advance of US D0.3 million for preparation of SCP which will come from the city's share of the A&OE funds and will be adjusted in the share of the city.

In the first year, Government proposes to give US D 29.85 million to each selected Smart City to create a higher initial corpus. After deducting the US D0.3 million advances and A&OE share of the MoUD, each selected Smart City will be given US D28.96 million out of US D29.85 million in the first year followed by US D14.63 million out of US D14.93 million every year for the next three years.

The focus of Government of India is wider than the EU focus on Energy, Transport/mobility and ICT. In our country, there is an additional dimension of need for: (I) economic development (II) employment generation (III) adequate water and its efficient distribution (IV) treatment and disposal of waste. Government of India believes in promotion of investments in Smart Cities, introduction of alternate approaches to investment and growth management by public-private partnership. It is also expected to empower city managers to prioritize, focus on issues affecting quality of life, and empower citizens and the private sector to partner. Smart Cities would deliver more effective decision-making, efficient governance and more appropriate and innovative investment vehicles, leveraging on global experience, competencies and success of Indian ICT sector. Government of India has further clarified that the budget is to either fund conceptualization of new project/scheme or retrofit in existing projects or locations; it is not for total project capital grant. These projects are expected to have clear revenue streams, suitable for non-recourse funding. The city level Pre-take-off plans in respect of quite a few cities are stimulating investment by engaging identified stakeholders i.e. ICT businesses, real estate developers, infrastructure providers, retail business and transport operators to support Public Private Partnership (PPP) projects and actions. The objective of Indian Smart City initiatives is to increase the resilience of the city. It is proposed to develop the 100 Smart Cities as satellite towns of larger cities by modernizing existing mid-sized cities.

The aim is to promote cities that provide core infrastructure and give a decent quality of life to its citizens, a clean and sustainable environment and application of 'Smart' Solutions. The focus is on sustainable and inclusive development and the idea is to look at compact areas, create a replicable model which will act like a light house to other aspiring cities. Smart Cities mission of the Government is meant to set examples that can be replicated both within and outside the Smart City, catalyzing the creation of similar Smart Cities in various regions and parts of the country. Area-based development will transform existing areas (retrofit and redevelop) including slums, into better planned ones, thereby improving livability of the whole City. New areas (greenfield) will be developed around cities in order to accommodate the expanding population in urban areas. Application of Smart Solutions will enable cities to use technology, information and data to improve infrastructure and services. Comprehensive development in this way will improve quality of life, create employment and enhance incomes for all, especially the poor and the disadvantaged, leading to inclusive cities. The GIFT city on the outskirts of Gandhi nagar and the Lavasa Hill City are examples of Smart Cities implanted in India as green field developments on a clean slate.

### 9.2.1.1 The journey so far

Under the scheme of things the cities selected under this mission for the Smart City projects are expected to work on incorporating the following features:

Promoting mixed land use in area based developments–planning for 'unplanned areas' containing a range of compatible activities and land uses close to one another in order to make land use more efficient. The States will enable some flexibility in land use and building bye-laws to adapt to change;

Housing and inclusiveness – expand housing opportunities for all;

Creating walkable localities – reduce congestion, air pollution and resource depletion, boost local economy, promote interactions and ensure security. The road network is created or refurbished not only for vehicles and public transport, but also for pedestrians and cyclists, and necessary administrative services are offered within walking or cycling distance;

Preserving and developing open spaces – parks, playgrounds, and recreational spaces in order to enhance the quality of life of citizens, reduce the urban heat effects in areas and generally promote eco-balance;

Promoting a variety of transport options – Transit Oriented Development (TOD), public transport and last mile para-transport connectivity;

Making governance citizen-friendly and cost effective – increasingly rely on online services to bring about accountability and transparency, especially using mobiles to reduce cost of services and providing services without having to go to municipal offices. Towards this objective steps such as forming e-groups to listen to people, obtaining feedback, using online monitoring of programs and activities with the aid of cyber tour of worksites have been thought of.

Giving an identity to the city based on its main economic activity, such as local cuisine, health, education, arts and craft, culture, sports goods, furniture, hosiery, textile, dairy, etc.

Applying Smart Solutions to provision of infrastructure and services for area-based development is envisaged. For an area-based development is envisaged for example, making areas less vulnerable to disasters, using fewer resources, and providing cheaper services. The Central Government will chip in US D 74.63 million – US D14.93 million per year, a matching contribution is expected from the respective State Government/ULB concerned. Other sources of financing the Smart City projects include user fees, revenue from Public-Private Partnerships, monetization of land, sale of development rights, municipal bonds, borrowings from bilateral and multilaterals, tapping National Investment and Infrastructure Fund (NIIF), convergence with other Government schemes, property tax, profession tax, entertainment tax, advertisement tax, parking charges and Octroi/entry tax.

As per the Govt. of India's MoUD'sorder bearing no. K-15016/10/SC-15/Part II dated 26th June 2015, each municipal corporation selected for Smart City project needs to set up an advisory board having NGOs and other stake holder associations so as to get feedback from the people at large. A special purpose vehicle is to be set up with participation of the local/planning authority, in the form of a company to look after all aspects of the Smart City project. The Chief Executive Officer (CEO) of such a company shall be the convener of the advisory body and could in most cases be, the municipal commissioner. The SPV will be a limited company incorporated under the Companies Act, 2013 at the city-level, in which the State/UT and the ULB will be the promoters having 50:50 equity shareholdings. The private sector or financial institutions could be considered for taking equity stake in the SPV, provided the shareholding pattern of 50:50 of the State/UT and the ULB is maintained and the State/UT and the ULB together have majority shareholding and control of the SPV [27].

The Bureau of Indian Standards (BIS) have laid down the parameters and the thresholds of various physical and social infrastructure service components to be achieved at the end of the 5 year implementation period envisaged for meeting the targets by going in the mission mode, vide its notification dated 30th Sept. 2016. Suggestions and objections on the draft standards were to be submitted latest by 30th November 2016. Further developments in this regard since then are not known.

### 9.2.1.2 Strategies employed

Broadly following strategies have been/are being considered for implementing the Smart City projects by various city governments for area based and/or pan city development:

Judicious Land use planning, development and zoning regulations to facilitate and promote desirable outcomes with due thought towards applications of ICT in spatial planning.

Regional Urban System Planning Approach to plan balanced urban hierarchical structure of settlements, avoid concentration of population and restrict speculative tendencies.

Replacement of traditional master planning approach with strategic planning approach, identification of priority zones and discouraging low-rise-low-density and high-rise-high density development are a few other strategies.

Promote Transit Oriented Development with medium to high density mixed development along rapid transit corridors. Induce high speed, high capacity, environment friendly public transport system like metro, monorail, BRTS. Develop an optimum mix of uses that reduces peak crowding and spreads travel demand throughout the day. Pedestrian focused development near transit terminals needs to be encouraged.

Promote use of specially designed bicycles to reduce carbon footprint and improve citizens' health. (Figure 9.5) Use Smart technology for user friendly interface.

**Figure 9.5** Promoting Cycling [35].

Liberalization policies, ineffective urban planning protocol and poor urban governance detrimental to urban land management need to be overhauled completely.

Different standards for Brown field and Green field situations will have to be adopted. Up gradation of infrastructure will have to be carried in the brown field case duly regarding space and technical feasibility aspects. Beyond 25% could be rather unrealistic. Green field planning should aim at 150% of present norms for parking, open spaces, utilities provision.

E-Governance, online payments, user friendly interface creation.

Creation of Intelligent buildings, city systems should be encouraged. Efforts made towards energy efficiency, greener/eco-friendly systems should be rewarded. Conjunctive use of available resources such as built space – in terms of time of the day, types of activities should be thought of seriously.

Standards to be derived for all aspects like – Environment, socio-economic living aspects i.e. smart people, urban mobility, technology enablement, housing for all, inclusiveness of all sections of society – physically challenged, economically disadvantaged, senior citizens, children, youth, females, reserved categories, physical infrastructure provisioning, aesthetically pleasing environment.

Urban design controls, built spaces utilization controls need to be put in place for creating an appealing ambience.

Promoting Internet of Things (IoT) is a good idea. (Figure 9.6) The Internet of Things (IoT) describes a system where items in the physical world, and sensors within or attached to these items, are connected to the Internet through wireless or wired network connections.

## IoT scope

The interactions between entities like sensors, connectivity, people and processes are creating new types of smart applications and services. (Figure 9.7) IoT will connect everything from industrial equipment to everyday objects that range from medical devices to automobiles to utility meters. Internet of Everything" (IoE), brings together people, process, data, and things to make networked connections more relevant and valuable than ever before by turning information into action (Lopez Research, 2013). There are many ways that IoT can help governments build smarter cities. The use of Smart Computing technologies will make the critical infrastructure components and services of a city – which include city administration, education, healthcare, public safety, real estate, transportation, and utilities – more intelligent, interconnected, and efficient. Availability of IPv6 protocol now should ease the

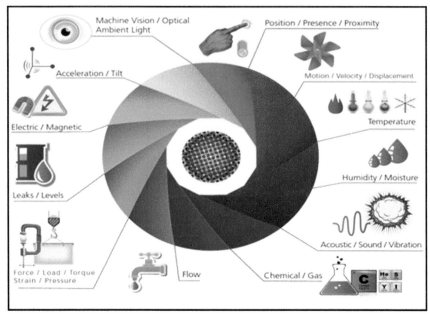

**Figure 9.6**   IoT [40].

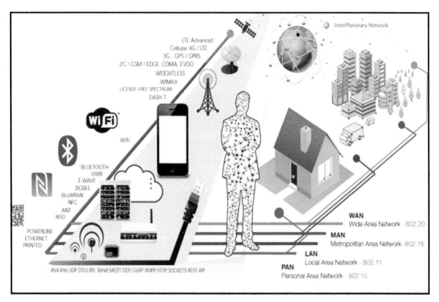

**Figure 9.7**   IoT Applications [40].

connecting task multifold. Economical provisioning wherever feasible should be made.

## 9.2.2 Outcome So Far

So far in 3 phases the Urban Development Ministry of the Government of India has cleared 60 cities to be taken up for implementation of the Smart City Project during the 5 year period since the approval granted in 2016 as per the MoUD's mandate. These cities will impact the lives of 72.2 million people with a total outlay of US D19666 million. This includes Pan City solutions costing US D3901 million and the rest on Area-based Programs. Special purpose vehicles have been incorporated in respect of 53 cities so far out of the 60 cities mentioned above [28]. Detailed Project Reports of proposals of many cities are at different stages of preparation/approval. These are being prepared by the Municipal Corporations/Special Purpose Companies set up by them either in-house by engaging existing regular employees or specially hired Planning professionals and managerial personnel or by engaging consultancy firms offering a host of services [29].

These 60 Smart Cities have visualized 329 projects categorized into 78 different initiatives aimed mostly at Area-based Development. The Area-based Development Strategies being adopted by the 13 fast track winning cities include both – redevelopment and retrofitting measures. Redevelopment initiatives comprised underutilized buildings, dilapidated buildings, reshaping and integration of open spaces. Whereas retrofitting encompassed transit infrastructure areas, heritage precincts, central business districts, markets, street façade improvement, flood control measures, lake/sea shore/river precincts, etc. Under the Area-based Development component some cities have taken up unique projects like city history museum, incubation centre, start up, skill development centre, interactive history museum, rental housing for construction workers/general public, GIS based land/property management system etc. Even mega projects like stadium, convention centre have been conceptualized.

Further the top 20 cities in the final ranking have together developed/are developing 29 different applications as part of e-governance with an allocation of US D 131 million [30]. The experience so far as reported in the press recently with regard to the implementation of the Smart City project components through private sector participation has been rather slow, except in case of some 3 cities led by Bhopal [31].

## 9.2.2.1  Urban planning, development and management: a continuous process

If we look at the entire realm of Urban Planning, Development and Management from the perspective of an ever evolving continuous process, then we realize that our cities have been instrumental in providing their residents with better facilities over time. Even before the formal launch of the Smart City program, taking cue from the Central government's reform measures envisaged in the JNNURM, a lot of Urban local bodies have adopted and are trying to implement modules relating to eight basic services such as property tax, accounting, water supply and other utilities, birth and death registration, grievance monitoring, personnel management, building plan approval and health programs. These innovative city strategies are aimed at providing user friendly interface to the citizens, digitalizing the records, easy access to information, cashless transactions, transparency in the dealings, consultative approach, stakeholders' say in matters of public interest, preventive actions to check leakages and thefts, etc. thereby managing the delivery of services. A glimpse of such efforts made by different city governments is presented here [4, 31]:

Coimbatore:
Computerized building plan approval scheme involves online submission of all requisite documents like ownership of land, applications for NOCs from various departments, payment of scrutiny fees, development charges, other levies, etc. building plans, certification by architect etc. and approval thereto by automated procedures in tune with development plan proposals and building by-laws governing sanctions, in a specified time line [32].

Jamshedpur:
Utilities Company providing an IT enabled 24/7 single window call centre and customer database to recycle around 40 million liters of sewage water in addition to the 30 million liters of water being treated currently. And in the next one year, it is possible that JUSCO could reuse 40 million liters of recycled water per day for industrial and gardening purposes in the city [33].

Surat:
CCTV cameras have been installed all around the city to control traffic and safety. SCADA operated waste water & water treatment plant for quality management.

Ahmedabad:
The city govt. efficiently runs the Bus Rapid Transit System on several routes, enabled by the GPS to transmit real time information of arrival and departure to the passengers (Figure 9.8) [34].

**Figure 9.8**  Bus Rapid Transit System in Ahmedabad [34].

Bangalore:
Has Geographic Information Systems (GIS) to operate property tax administration.

Kanpur:
The Municipal Corporation uses GIS for controlling municipal revenues. The initiative involved creation of a GIS-based spatial property database using high resolution satellite imagery, geo-referenced control points taken by dual-band GPS (DGPS) survey, and updating the data by ground truthing (validation). The property boundary layer was then migrated on top of this GIS base map to create homogenous GIS data layers. Tax Assessment data preparation using existing property details and door-to-door contact surveys followed with re-numbering of properties to assign a unique identifier to each property and fixing of house numbers. Development of desktop application for property database repository management and tax assessment led to an informative and interactive web-GIS system for online property tax calculation and payment [35].

Hyderabad:
Is using GPS ad GPRS technologies to cover solid waste management and for maintaining parks and street lighting.

Mumbai:
With space constraints, creating a garden in Mumbai to grow pesticide-free vegetables and fruits may seem like a far-fetched idea. But one group of organic farming enthusiasts has shown how growing an organic kitchen garden in the city is quite an achievable feat. Soil created using organic waste is not inert. It is full of microbial life, which can be used to grow organic food without having to use synthetic fertilizers. The concept is food should be grown where waste is generated and being decomposed. The idea is to utilize biodegradable kitchen waste instead of sending it to landfills. Of the total garbage sent to landfills – around 8,000 metric tons – in Mumbai, 40% is organic and can be used as manure. Terrace farming has set up a model for young citizens for developing environment friendly technology. The methodology involves building and using AmrutMitti (a nutrient-rich soil), Seed sowing and transplanting of seedlings, Pruning for fruits and vegetable plants to create a better canopy and make them more productive, Building trellises for creepers and adopting techniques to harvest maximum light and maintain high biodiversity in farms/terrace gardens [36].

Delhi:
In Delhi a proposal was made for "Waste mining" to use the solid waste to produce energy. The 5 Municipal Corporations situated in the NCT of Delhi together operate 3 Waste to Energy Plants at Timarpur-Okhla (1950 TPD, 16 MW), Ghazipur (1300 TPD 12 MW) and Narela (3000 TPD, 24 MW) [37].

Lavasa:
It is touted as India's first e-city., My City Technology – a joint venture set-up by Lavasa Corporation and Wipro would help in city management services, e-governance, ICT infrastructure and value-added services, including proposing and implementing intelligent home solutions and digital lifestyles for the Lavasa citizens. (Figure 9.9) Lavasa homes will offer touch-point automation, occupancy-based lighting, door and motion sensors, beam detectors and on-call transport services.

GIFT:
GIFT city coming up in Gandhi Nagar, Gujarat, will have a central command centre to monitor the city-wide IT network and respond quickly during emergencies, energy-efficient cooling systems instead of air conditioning, and high-tech waste collection systems. Cars will remain outside, and there will be moving walkways to get to the city centre (Figure 9.10).

**Figure 9.9** Lavasa hill station [38].

**Figure 9.10** Artist's impression of GIFT city [38].

Navi Mumbai:

To offer Navi Mumbai residents an enriching lifestyle replete with all the necessities of modern urban living, CIDCO has strived hard for over 4 decades now. Development of facilities like Vishnudas Bhave Auditorium (Figure 9.11), Central Park (Figure 9.12) and NRI Housing Complex (Figure 9.13) are some of them. Moreover as a commitment towards inclusivity, initiatives like schemes covering plots for Senior Citizens' facilities, plots for special schools, 3% reservation in CIDCO built tenements, barrier-free environment for the people with disabilities, plots as well as built up

**Figure 9.11**    Vishnudas Bhave auditorium [39].

**Figure 9.12**    Central park [39].

spaces to NGOs, women's self-help groups at concessional rates, plots for day care centers, for project affected persons preference in house allotment, social facility plots,  stipends, scholarships for children, infrastructure up gradation in villages, land reclamation work tenders, over and above the land

**Figure 9.13** NRI housing complex [39].

acquisition cost, developed plot, etc. 85 sites for markets with basic amenities for 5500 hawkers, platform for artisans to meet face to face with prospective buyers by way of the Urban Haat project, etc. have been taken. Substantial outlay for Affordable Housing (60000 units – US D1597 million), Railway & Metro Projects – US D1908 million, Infrastructure Development Projects – US D1221.5 million, Pushpak Nagar – Greenfield Development – 230 Ha Project Cost: US D140.5 million has been proposed. Other measures include Vigilance Cell to deter irregularities, Computerized Online Grievance Redressal System, Installation of 430 CCTV cameras, Sensors, Citizen centric portals, Disaster management response systems, Allotment of plots to PAPs by computer based lottery, E tendering for projects costing more than US D4478, Auto DCR for scrutiny of building permission cases, GIS system and SAP system, Single window clearance, e-payment facilities, information kiosks, Smart City Chair In National Institute of Urban affairs, New Delhi [38, 39].

### 9.2.3 Observations

Our cities are not run as a system. Therefore they are less efficient in distribution of resources and infrastructure in a just and equitable manner. There are overlaps of roles and responsibilities and most of the time they are not well-defined leading to either duplication of efforts resulting in lower productivity or no efforts due to confusing command structure. Today there is political leadership at ward level undermining the city administration at middle

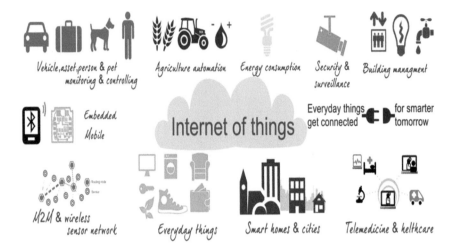

**Figure 9.14**    Internet holds promise for the future in city governance [40].

management level. The city administration top management has the power to act but is entangled in centuries old procedures, rules and regulations. The city mayors in most of the cities are not directly elected, have short tenure of around a year and the position is more ceremonial than real seat of power. As a result the administration either plays safe or undermines the system. Property records though computerized are dated (not updated dynamically) and existing Internet of Things (IoT) applications (Figure 9.14) are just computerization of dated administrative procedure with legacy of past rules and regulations. Today there is a shortage of trained work force to play the role of "Development Manager", working on bankable catalytic projects. These projects also require commitment of City Managers.

We need to take note of the present scenario of multiplicity of agencies operating in India attending the same issues – many a times with conflicting interests or lack of coordination amongst them. There is a need to remedy this situation so that a favorable environment is created to reap the benefits of the latest smart technologies, IT solutions, web-based services, sector advances in transport, etc.

## 9.3 Conclusion

The city governments will thrive to remove the lacunae in the systems and engage in course correction in a bid to meet the expectations of its citizens in an atmosphere of competition amongst the peers. The infusion of new ideas germinating from various corners of India by way of brain storming sessions to devise better strategies will have a positive outcome. The city administration will aim at delivery of services in a transparent and efficient way to improve its image in the eyes of the stake holders concerned. The consultative and public participatory process adopted to know the pulse of the people has set the priorities right for delivery of services. The new initiative of the Government of India to make cities capable of facing the twenty-first century challenges with the generous reliance on the latest frontiers of technologies available, serious efforts towards capacity building of the personnel manning the projects and above all employing methods to make them financially sustainable with little aid will go a long way in creating a Resurgent Urban India.

The importance of deploying advanced technologies for solving the problems of city governance with ease cannot be undermined. But at the same time while embracing newer and smarter technologies to offer ease of doing business, care should be taken to make available services to the lowest denominator of the society too, in a way he/she is comfortable with. Suffices to say that at any given point of time concurrently 3 generations of technology platforms should be made available to ensure the city maintains the "inclusivity" characteristics and no person remains isolated from the main stream of city life. In short, both online and offline modes should be offered to the end users. Technology should be looked at as a tool to achieve certain objectives and not as an end in itself. Else the city fabric will lose the human face forever!

## References

[1] Schirber, M. (2005). Available at: https://www.livescience.com/6893-cities-cover-earth-realized (accessed on 11 March).

[2] ITU-T FG-SSC. (2014). *ITU-T FG-SSC Technical Report on 'Smart Sustainable Cities: An analysis of definitions,* p. 14.

[3] Geospatial Media and Communications and United Nations Economic Commissions for Europe (2014). "Land information systems for smart cities," in *Geospatial world forum 2014*, Geneva, Switzerland. Available at: https://www.unece.org/fileadmin/DAM/hlm/wpla/workshops/geneva 2014/Land_Information_Systems_for_Smart_Cities_26122013.pdf

[4] United Nations. Available at: https://www.un.org/

[5] EBTC. (2014). *Report on Smart-Sustainable cities and Communities Initiative in India*, p. 8, 9. European Business and Technology Centre, Belgium.

[6] European Union. Available at: https://europa.eu/rapid/press-release_ MEMO-15-5712_en.htm

[7] Gov. UK. (2013). *Smart cities:background paper.* GOV. UK, UK: Government Department for Business, Innovation and Skills, 135.

[8] Patel, C. (2015)."Smart cities: issues and challenges in Indian context," in *SVNIT at the National Conference on Sustainable Smart Cities in 2015*, SVNIT, Surat

[9] Available at: https://www.iadb.org/handle/11319/7727

[10] Available at: https://www.urbantransform.eu/wp-content/uploads/sites/ 2/2015/07/amsterdam-becoming-a-smart-city.pdf

[11] Available at: https://www.itu.int/dms-pub/itu-t/oth/23/01/T2301000019 0001PDFE.pdf

[12] Available at: https://www.lta.gov.sg/content/ltaweb/en/industry-matters/ traffic-info-service-providers/real-time-traffic-information.html

[13] Available at: https://www.weforum.org/agenda/2016/06/dubai-is-build ing-the-worlds-biggest-solar-power-plant/

[14] Available at: https://www.uaeinteract.com/news/default3.asp?ID=231

[15] Available at: https://india.gov.in/spotlight/smart-cities-mission-step-to wards-smart-india

[16] Kumar, E. (2015). "Urban Nexus and Linkages to GOI's New Schemes," in *5th Regional Workshop on Integrated Resource Management in Asian Cities: The Urban Nexus*, Chiang Mai, 17–19 June, 2015. Available at: http://www.unescap.org/sites/default/files/Session%207%20-%20ICLEI %20SAS-%20GoI%20policies.pdf

[17] Census of India 2011

[18] Available at: https://www.data.worldbank.org/country/india

[19] The Challenges of Urbanization in India, 12th Plan, Planning Commission of India, 1–5.

[20] Available at: https://en.wikipedia.org/wiki/List_of_countries_by_GDP (PPP)

[21] Ministry of Urban Development. (2014). *Draft Concept on Smart City Scheme*, 1–46 (in preparation).

[22] BSI: 180, 2014) (2c); Available at: https://www.researchgate.net/publica tion/2667764451-Smart-Cities-Contradicting-definitions

[23] Available at: https://www.smartcities.gov.in/content/innerpage/whatisa smartcity.php

[24] Available at: https://www.smartcities.gov.in/content/innerpage/smart-city-features.php

[25] Available at: https://mhupa.gov.in/writereaddata/Mission%20Overview %20English.pdf

[26] Available at: https://www.smartcities.gov.in/content/innerpage/spv.php

[27] Ministry of Urban Development. (2015). *Smart Cities Mission-Transform-Nation.* Mission Statement and Guidelines, pp. 12, 35–36, Ministry of Urban Development, Government of India.

[28] Available at: https://www.smartnet.niua.org/smart-cities-network

[29] Available at: https://www.smartnet.niua.org/smart-cities-network

[30] Available at: https://www.smartnet.niua.org/smart-cities-network

[31] Press report in Maharashtra Times, Mumbai 19/01/2017

[32] Available at: www.ccmc.gov.in/ccmc/index.php/services/town-planning/ 35-toplink

[33] Available at: www.india.smartcitiescouncil/article/jamshedpur-first-zero-sewerage

[34] Available at: www.ahmedabadcity.gov.in/portal/jsp/static-pages

[35] Available at: https://www.kmc.up.nic.in/smartcity.htm

[36] Available at: https://www.hindustantimes.com/mumbai/mumbai-organic -is-the-way-to-grow-for-these-urban-farmers/story-GeRZUmBrnbpyUL vxcPG0QM.html

[37] Available at: https://www.delhi.gov.in/wps/wcm/connect/environment/ Environment/Home/Environmental+Issues/Waste+Management

[38] Lele, M. D., Mane, P., and Gawde, D. (2015). "Inclusive Smart Cities – Case study of Navi Mumbai," in *National Town & Country Planners' Congress*, Chennai.

[39] Available at: https://www.cidco.maharashtra.gov.in

[40] Presentation on "Understanding Internet of Things (IOT) and Smart Cities" by Dr. Gayatri Doctor, Asso. Professor, CEPT University, Ahmedabad at the National Conference on Sustainable Smart Cities in 2015, at SVNIT, Surat

## Biography

**M. D. Lele**, Retired as Chief Planner of the City & Industrial Development Corporation of Maharashtra (CIDCO), Navi Mumbai and is currently working as an Associate Professor at the Vishwa Niketan Institute of Management, Entrepreneurship, Engineering & Technology, Mumbai, India.

# 10

# Smart Cities and Business Model Ecosystems

Peter Lindgren and Ramjee Prasad

Department of Business Development and Technology, Aarhus University, Herning, Denmark

## Abstract

The Smart City Business Model Ecosystems (SCBMES) seems to be established and approached very differently from one Smart City to another on the Global Scene. Those Smart City terminologies we know of today seems to be very much focused on one fits all – but this may not open up for the real potentials and even several hidden potentials of the Smart City Concept. Smart Cities may be claimed to become one of the futures most important approaches to grow out society's economy – creating new types of valuable BMES. We show and argue however – in this paper – that the Smart City concept combined with the BMES approach may potentially result in even more valuable and sustainable outputs of Smart Cities. If we are able to "see" and "sense" – "act" and "do" – and value the Smart City BMES with interdisciplinary competences – the Smart Cities BMES around the global world might be able to release much more potential than initial expected. This demands however that we begin to bring people and Business really into "the concept of Smart Cities". Today – we claim on the basis of our research – that many SCBMES have not really brought the people and the businesses into the so-called Smart Cities – and this have left several Smart Cities and Smart City projects in deep challenge – even made them become "Ghost Cities" – which could be argued is not really "Smart" and "intelligent".

**Keywords:** Smart Cities, Business Model Ecosystem, SCBMES, Business Model innovation.

*Breakthroughs in Smart City Implementation,* 243–286.

## 10.1 Introduction to Smart City and BMES

Smart Cities seen "just" – as something related to physical, geography and "technology" might in the future be too narrow, not "smart" and "intelligent" enough a view and "thinking" to the Smart City approach. Context boarders and individual approaches to view the smart city differently, in an interdisciplinary, sustainable and business model ecosystem context have initially had hard times to gain the Smart City agenda but when adapted might release new and valuable potentials to our society. Smart city projects have originally been framed

> "*as technology/R&D projects: testing and delivering promising new technological solutions for sustainability, pressing societal problems, better public spaces, or improved public services. But smart city technology, like any technology, is not neutral or independent: it intermingles, in complex ways, with people and organizations that use it, reject it, or embrace it. It has ethical implications, and sometimes unintended or undesired outcomes, and it creates conflicts of power and interests*" [75].

Back in the early and mid-2000s, cities everywhere wanted to turn "digital" or "intelligent." Especially in Western Europe and North America – and now also in many cities in China, India, Middle east, South America and other places around the world, municipal and governments drafted and still drafts strategies to keep up with modern technology, digitalize communication with their citizens and introduce high-tech information exchange [63, 64]. The advent of smartphones and near-universal WiFi connections scupper many of these traditionally top-down approaches, but also opened up new avenues – and some would say new "rawmaterials" for business model innovation. This evolution have now been highly adapted into Europe [17, 18, 61], US, India [21, 22, 42, 60], China [10], Korea, South America and many other places around the world.

European Union e.g invested Euro 200m in the European Innovation Partnership for Smart Cities and Communities [72] to adopt the Partnership's Strategic Implementation Plan (SIP) from 2017–2019, with the aim to serve as the basis for speeding up the deployment of Smart City solutions in Europe. The Strategic Implementation Plan behind this Smart City initiative sets out a broad range of new actions and approaches to encourage European cities to become smarter. The plan concentrates on how to drive forward improvement in **buildings and planning**, **new Information Technologies**,

**transport and energy**, and **new ways of integrating these areas**. The plan also suggests **improvements to the way that cities are run** with **better ways of involving citizens and more collaborative ways of doing things**. It suggests **innovation zones**, **new business models**, a **re-evaluation of rules and legislation** and a **more standardized approach to data collection and use**, to **enable better comparisons between approaches and between cities**. This is just the beginning of a large scale programme of work in EU including a manifold of partners. An important part of that work will be the "Lighthouse Projects" – cities, which will demonstrate and deliver Smart City solutions on a large scale. These Smart City projects will partly be financed by the European Commission's Horizon 2020 Research Funds, together with business and public funding – to help spread these new solutions to other cities. By Economies of scale EU hope to help the many "innovative" and "high tech" solutions to become "the norm" – and available more easily to all cities and neighborhoods in EU. The SIP is drafted by a great variety of actors from businesses, cities, civil society and research, focusing on three specific areas: **sustainable districts, sustainable urban mobility, and integrated infrastructures across energy, ICT and transport, and proposes a variety of actions to drive forward improvements in these areas** [73].

Winden et al. [75] presented in May 2017 the wide context of Smart City projects, addressing non-technological aspects of smart city projects. This was one of the first attempts on Smart City definitions and approaches bringing and relating the human more to the Smart City concept. Further Wilden address also different other facets such as the difficulties to make businesses with different agendas collaborate with each other. He address the involvement of different stakeholders on how to divide or share the returns and outputs of Smart Cities and Smart City BMES (SCBMES). Risk related to these issues, were also addressed in Windens research [74]. The analysis showed hereby some of the non-technological aspects and challenges of smart city projects (partnerships, business models, scaling potential) and concluded that Smart City or SCBMES are not just about developing and applying technology but demands new networking, new management and new Business Model Innovation (BMI) competencies. The research verify that Smart City Business Models and SCBMES are not and cannot be developed and implemented by one single business. Smart Cities but can only really take shape in networks of Businesses and with involving of the humans – citizens/end users in a sustainable network of business models – a SCBMES or a network of SCBMES. Interdiciplinary competences and Partnerships are necessary to form the SCBMES of the future and somebody has to motivate and make them work together – in often a very complex setup and BMI room – where all are given "room" and "space" to work and work together. This with respect to the fact that each participant can work differently, "see" differently, "sense" differently – within the aim of "facing" and solving the different challenges of a SCBMES.

A strategic understanding of smart cities combined with the BMES approach could be a new business model innovation leadership (BMIL) approach to SCBMES solutions – adding new dimensions to Smart City Concept related to previous terms and "thought worlds" of smart cities.

## 10.2 Research Metrology and Design

The research for this paper was carried out on behalf of a literature study of Smart Cities and Smart City approaches combined with secondary case examples from material and research on Smart Cities – from US, EU, China, India, Korea, Middle East Asia and South America on Smart Cities. Amsterdam University of Applied Sciences provided us with a systematically case analysis of 12 smart city projects in Amsterdam. This gave us the basis for this paper. Parallel to this we observed a large Smart City project in Aarhus Municipality, Denmark funded by the European Commision.

The aim of the paper is to investigate how a combination of the Smart City approach and the BMES approach can value and increase our understanding and development of future Smart city BMES and Business Models – in the favour of the users, customers, people and businesses of Smart Cities.

## 10.3 Theoretical Background

To date, there is no internationally recognized definition of a smart city – but most definitions are related to a combination of "smart", "intelligent" solutions of technologies – using technology to "smarten up", change or even create a city's e.g. on energy supply, buildings, mobility options, healthcare, governance and education. Often the aim is to achieve more efficient and better integrated infrastructure system of cities. Although these parameters are often used by international rankings [17, 18] to measure a cities status and progress towards greater "smartness," "more intelligence" the cities themselves often view and approach the concept differently. Often the cities have different visions, goals and strategies to Smart City as can be seen later on in this paper.

"The definition of smart cities was initially driven by the big technology businesses such as IBM, Cisco and Philips. Digital innovation was to be linked with "intelligent" and smart urban projects. In many definitions, the word 'smart' referred – and still does – to the application of new technologies to improve urban services or the quality of life of the city.

Many of the Smart City project in the past and also even today are driven and based solely on technology – and advanced technology. Barcelona's smart city focuses on e.g. **opening up technology to the citizen and investing in innovation and business via the new technology**, whereas Rio de Janeiro is focusing on becoming a smart city of **security via using ICT.** "There is no single, all-encompassing definition," Ginzler [61]. – "Every city is different and needs to rely on its individual strengths and tackle the weaknesses." A very important statement seen in the context of combining Smart City with BMES. Bakici, Almirall, & Wareham (2013) [6] defined smart cities as "cities that **utilize information and communication technologies** with the aim **to increase life quality** of **their inhabitants** while providing sustainable development". However it is more blurred to see what sustainable development actually means to whom, to how, to which stakeholder.

ICT plays without question in most definitions of smart cities a pivotal role in transforming and creating a city more adapted to the contemporary needs of its citizens – a smart city "embedded with ICT" to those and for those living and operating in the cities. However the business and the business models seems in the very early approaches of Smart City to be left out of "the definition". The human and people in the Smart City was mainly used as a kind of "media" and "given thing" that would like and accept in any case the new technologies and the output and spin out of the technology.

Other definitions of smart cities began to include also some reference to the use and amount of use of ICT for making modern cities more suited to the needs of citizens (Chourabi, et al., 2012) [11] – the humans. Many began to view cities as smart **when "investments in humans, social capital and traditional transportation resulted in modern (ICT-based) infrastructure, sustainable economic growth and a higher quality of life of the inhabitants in the smart city. This with a wise management of the Smart Cities natural resources, through participatory government".**

Some developed the Smart City concept by arguing that a city is smart when it manages to connect the physical infrastructure, the IT infrastructure, the social infrastructure, and the business infrastructure to leverage the collective intelligence of the city. ICT takes here again "center stage" in the definition – as many other definitions of smart city and the development of a smart city have done until now. ICT is regarded as a clear key driver of smart city initiatives – few focused however initially on the humans living inside and outside the so called Smart Cities. ICT is seen as an essential "building block" that a smart city needs in order to make **the city smart**.

Some Smart City projects and approaches like e.g. The Amsterdam Smart City (ASC) platform toke however a broader perspective to the Smart City concept by including projects without a strong technology component and focus alone. It 'supports innovative ideas and solutions for Smart Cities issues' and thus contributes to the livability of the urban, the SCBMES – meaning that it also address not just the SCBMES inside the Smart City but also BMES and environment outside the Smart City in focus. In this case also the impact of establishing a Smart City that is related to the BMES and environment around or outside the so called Smart City. The smart city projects and definition in this context address and increases the definition with three criteria not seen that much before:

1. development or use of new technology, intending to generate not only economic value but also ecological and/or social value
2. element of innovation or experimentation
3. partnerships, networks and networking

Vienna, Barcelona, Copenhagen, Aarhus and Amsterdam continued that journey of changing the Smart City approach by "cultivating" the Smart City Approach and definition, using technology as a tool rather than as an end in itself – to create an interconnected, user-friendly and human integrated city and urban life within the city.

To finalize the theoretical coverage of the ICT approach to Smart Cities 5 ICT essentials related to smart cities were proposed by – Escher Group [16]:

1. **Essential #1 Deployment of Broadband Networks** – *"to foster the development of a rich environment of broadband networks that support digital applications, ensuring that these networks are available throughout the city and to all citizens in the smart city"* [31]. This approach included a broadband infrastructure that combined cable, optical fibre, and wireless networks offering maximum connectivity and bandwidth to citizens and businesses located in the city". "Smart Cities could hereby use broadband wireless networks to enable a wide range of smart city applications that could enhance safety and security, improve efficiency of municipal services and promote a better quality of life for residents and visitors.

2. **Essential #2 Use of Smart Devices and Agents** – *"to ensure that the physical space and infrastructures of the Smart City are enriched with*

*embedded systems, smart devices, sensors, and actuators, offering real-time data management, alerts, and information processing for the city administration*". "The presence of these devices combined with wireless connectivity throughout a city facilitates a richer and more complex digital space within the Smart City, which in turn could potentially increase the collective embedded intelligence of a city. This collective embedded intelligence allows relevant stakeholders of the city to be informed about the city's physical environment and *facilitates* the deployment of advanced services like spatial intelligence. It also potentially paves the way for developing other innovative BMES that help to link the Smart City with its people and visitors through technology" [25]. The argument is that this will embed intelligence, created by the use of embedded systems and other ICT intensive solutions, and it will become the nervous system of modern economies through making cities smarter".

3. **Essential #3 Developing Smart Urban Spaces** – On behalf of Essentials 1 and 2 Escher group propose that it is possible – developing smart urban spaces, by connecting the embedded systems, sensors and smart devices located across the Smart City together with forming a cohesive and integrated ICT infrastructure for the smart city. Smart urban spaces are areas of a city that leverage ICT to deliver more efficient and sustainable services and infrastructures within that specific area. These smart urban spaces comprise a wide range of business model innovations that can be of enormous environmental and economic benefit to both the district and the city at large. Creation of applications, which enable data collection and processing, web-based collaboration, and actualization of the collective intelligence of citizens. The latest developments in 5G, cloud computing and the emerging IoT, open data, semantic web, and future media technologies are forecasted to have much to offer smart cities – and enables them to "look" and become even smarter. These technologies many authors claim will be able to assure economies of scale in infrastructure, standardization of applications, and turn-key solutions for software as a service (SaaS), which dramatically will decrease the development costs while accelerating the learning curve for effective functioning of smart cities [54].

4. **Essential #4 Developing Web-based Applications and e-Services** – "The availability of ubiquitous ICT infrastructures is claimed to be able

to stimulate the development of new services and applications by various types of users, and allow for the gathering of a more realistic assessment of users' perspectives. This is possible by conducting acceptability tests directly on the infrastructures already in place and functioning in the smart city. Smart City Living Lab networks [70, 71] are said to be able to help to make the testing of new applications and e-services easier and should be used as building blocks for the more efficient development of smart cities". – "Smart cities commonly deploy online services across different sectors of the city, for instance a city airport will require different e-services to a city hospital. Smart city e-services include services for the local economy and its development, tourism, the city environment, its energy and transport services, security services, education and health services and so on.

5. **Essential #5 Opening up Government Data** – "The effective use of government data can precipitate the smart evolution of a country's cities, creating national competitive advantage for the country in question. Greater openness of information, documents and datasets held by public bodies. The Right to Information movement, which promotes a public right of access to information from a human rights perspective. The Open Government Data movement, which uses predominantly social, ethical, economic arguments to encourage the opening up of government data. Putting such information into the public domain is said to be able to benefit society by creating conditions for more social inclusive service delivery and for more participatory democracy. Many authors argue that this can stimulate the economy by allowing the possibility for third parties (e.g. individuals, private enterprises, civil society organizations) to create new products and services using public data. However, they are not very precise, where the businesses is and what type of business models are potential – and where and to whom should the new products and services be sold. In other words many talks of the potential benefits and values of smart cities – but few verify and documents the businesses and business models – and if they are sustainable to both the smart city itself and to those outside the smart city" [16].

Smart City build on standard tools are said to have significant benefits, because it was said to have critical new resource for fueling changes in value creation (economic, social and political) of a city or region. However, the literature are not very precise on how to capture these value, how to deliver and receive these values and not least how to consume the values – do BMI and Business. There is indeed a strong lack of how do our society make business on the concept of Smart Cities. In business model literature and research it has been proven that it is seldom enough to a business – just have users – and many users. In a long term perspective [37, 38] somebody, someday or some way – we define them as customers [37, 38] – have to and must pay for the intelligence, services, accessibility of the SCBMES.

The OECD identified 5 benefits – values – to opening government data for a city, region or country – for many equivalent to having a smart city:

1. **Improving government accountability, transparency, responsiveness and democratic control**
2. **Promoting citizens self-empowerment, social participation and engagement**
3. **Building the next generation of empowered civil servants**
4. **Fostering innovation, efficiency and effectiveness in government services**
5. **Creating value for the wider economy**

The idea of this Smart City approach is that the government – the smart city – should **change according to the need for a city's governing body** to **engage with its citizens and listen to their needs** when developing the city – smart city – is a general theme in much of the latest Smart City literature. However not all governments and "Smart Cities" want – or have considered the meaning and power of – taking this next step of Smart Cities fully. Opening up government data to citizens, some argues will encourages good governance. This is regarded in some culture as good, interesting, sustainable – but in other societies regarded not. In other terms Good Governments are seen very differently in different cultures – and the different Smart City approaches has to or are forced to take this into considerations

Good governance according to the OECD's term, in turn, encourages **public trust and participation** that enables services to improve. However few and even not OECD discuss if there is alignment between all stakeholders wants, needs and demands of value on the above mentioned. Elsewise the Smart City will just become an illusion – "a fantasy" – without the basic and fundamentals of a network – trust, ownership, power, momentum, optimal network structure [76].

Smart cities concept have often been combined with or made equal to the term e-government. Some [55] have studied the challenges of key Smart City projects related to e-government, and found that stakeholders' **relations is one of the very critical and important factors** to determine success or failure of a Smart City and Smart City projects. In other words the humans – people – and the businesses relations to the Smart City. For improvement in the quality of life of the urban areas, Smart City is claimed to be the **solution towards growth and sustainable development** but management of Smart City has to consider the ownership, openness, structure, security of Smart City, when data and infrastructure of Smart City potentially becomes fully open for access – and to all. In this case humans, people of the city and businesses of the city are important users, customers, network partners and "humand resources and competences" of the Smart City.

However as many have claimed [64] the ease of life kind of luxuries always comes with security and safety risks. Almost **all cyber-physical systems are prone to cyber threats – in this case also Smart Cities – SCBMES**. A robust and secure smart city is the right and necessity of every citizen – elsewise there will be no smart city – and as a result no sustainable SCBMES [38] and related BM's. However Smart Cities face a paradox – as while the City become more "Smart" and "intelligent" the risk and impact of cyber threats increases – or have until now increased.

For many countries, Smart cities are claimed and believed to be **the future growth engines** of their economies, offering their populations greater opportunities for education, employment and prosperity. But, the negative effects of their growth and the urban evolution in general [77] are also seen today as firstly negative results inside the smart city BMES – like traffic congestion, informal settlements, urban stretch, environmental pollution, exploitation of resources and a major contribution to climate change, even movement of war and terror right into the smart cities as Standford Peace Innovation Lab indicates [20]. Secondly radical consequenses for the rural areas and environment around the Smart Cities like – braindrain, resource drain, old houses left behind without anybody and any ressources to remove them. Tremendous concern about and as a following result high investment in security is the consequence for Smart Cities. Efficient and intelligent technologies – such as persuasive technology and persuasive business models [36, 39] could hold the answer to many of these urban or smart city challenges – but until now they are unfortunately not researched enough on – and those solutions to meet the challenges available today are judged to be in the very initial stage of business model innovation [34, 36, 39].

The word "Smart" have been implied – "intelligence" – and in relation to Smart City it is in literature narrowed down in many cases to **the automation**

**in the cyber physical systems of the city** for **bringing quality and ease of life to its citizens in a cost and resource efficient manner**. Future population and environmental issues – we propose – should be tackled with the sustainable development of the Smart cities with long term vision and mission – but should also in this context consider other BMES [38]and the environment around and "outside" the Smart City. Because of continuous population growth – and the strong movement from rural areas to the cities [77], several problems and challenges have evolved for the Smart City development in developing countries like India and China – but also for European and US Smart Cities. However these problems and challenges are different from one Smart City to another and should therefore be seen in different contexts. In this matter the BMES framework could be helpful to find out "how to cut the cake" or where "to drag the knife in" to maximize the value, cost and ROI of SCBMES both money wise and relate to other values [35].

Many smart city initiatives and developments – especially in India [42] and China [10] focus on the **core infrastructure elements** which include **adequate and clean water supply, assured electricity supply provision, sanitation with consideration of solid waste management, health and education, sustainable green environment, efficient urban mobility and clean public transport, affordable housing for economically backward people, safety and security of citizens, and last but not the least, good governance with citizen participation**. These core infrastructure key elements are depicted in Figure 10.1 and seems to be addressing needs mostly at the lower part of the need pyramid of Maslows model [41].

Some general attributes have initially been proposed related to defining a **sustainable smart city development**.

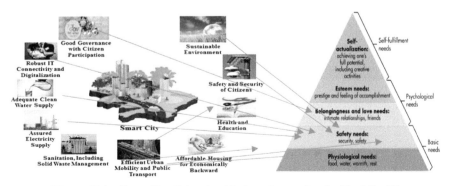

**Figure 10.1** Smart City Related to Maslow. *Source:* Inspired by [42, 41].

### 10.3.1  Area Based Development

New Cities should be planned with consideration of separate isolated green area and residential projects should be implemented at a distant place from parks and green zone. Similarly, business zones should be isolated as well and more greenery should be incorporated in the business infrastructure plans. Already existing city infrastructures should be modified with the retrofitting techniques. Smart applications should be introduced in the existing cities.

### 10.3.2  Housing and All-inclusiveness

Affordable housing should be made available to economically backward community. Green and clean houses for all is the need of smart cities. Building rooftops can be mounted with **solar cells to generate adequate amount of electricity** required by individual home. Smart buildings with ultra-new smart and secure features from ICT are the need of time.

### 10.3.3  Pedestrian and Bicycle Zones

For congestion reduction, separate walkable localities can be developed and maintained with surrounding greenery. On the same lines, bicycle side roads which are distant from main roads can be strictly developed and maintained. This should help in reducing the $CO_2$ emission in the city and also help in building healthy and sustainable environment.

### 10.3.4  Public Transport

Clean and timely public transport system with electric vehicles, separate lines for trams, buses and local trains are necessary. Usage of bio-fuels can help in **reduction of $CO_2$ emission**. Rapid transit systems will save lot of time and energy. Trains should be powered by electricity generated from renewable energy sources.

### 10.3.5  Green Energy

**Smart Grid** can bring **automation and uniformity in the electricity generation, transmission, distribution and usage**. Mini grids should be implanted for individual buildings. Solar and wind farms can provide the major amount of electricity needed for individual city.

### 10.3.6 Waste Management

**Recycle treatment plants** for city's **waste which can generate energy** in turn is required. Centralized mechanical system should be developed for collection of dry and wet waste. Waste food can be utilized in the **production of bio-gases**.

### 10.3.7 Health, Education

Ecologically supported ecosystems reduce the pollution and health of the citizens can be maintained well. Also a centralized multi-speciality hospital facility with minor costs is the need of every smart city. For the comprehensive smart city development, citizen participation is very much essential. State of the art education techniques should be able to develop "smart people" who can participate well in governance activities.

### 10.3.8 Security and Safety

**Security of various cyber physical systems and safety of the individual citizen** especially elderly people, women and children are of prime importance. The information collected from various sources face lot of security issues such as rights, duties and risks, etc. Cyber security mainly focuses on computing systems, data exchange media, and the actual information being processed.

ITU [13] have also tried to defined the Smart City concept as:
"A smart sustainable city (SSC) is **an innovative city that uses information and communication technologies (ICTs) and other means to improve quality of life, efficiency of urban operation and services, and competitiveness, while ensuring that it meets the needs of present and future generations with respect to economic, social and environmental aspects**"[25].

European Union [17, 18] defines smart city as:
"A smart city uses information and communications technology (ICT) to enhance its liveability, workability and sustainability".

Three parts to that job are specified: collecting, communicating and – crunching.

1. First, a smart city **collects information about itself through sensors, other devices and existing systems**.
2. Next, it communicates that **data using wired or wireless networks**.
3. Third, it – **crunches (analyzes) that data to understand what's happening now and what's likely to happen next**.

**Figure 10.2**   Smart Cities in European Union. *Source:* [17, 18].

The first model is wherein the existing physical systems will be retooled with digital infrastructure. Related to this example could be mentioned Aarhus Smart City – [70] http://www.smartaarhus.eu/ and Copenhagen Smart City – Initiative – [71] http://www.copcap.com/set-up-a-business/key-sectors/smart-city. The second model is redevelopment of the existing infrastructure by con-sidering reconstruction. Related to this examples could be mentioned Smart Cities like Shanghai, China, [10], Mumbai India [42, 44], Rio, Brazil [15]. The third model in greenfield approach, which considers the approach of building cities with completely new smart solutions. Related to greenfield SmartCity BMES in china is a good example [10] http://www.smartcityexpo.net/en/ – http://nextshark.com/china-building-first-smart-city-not-beijing-shanghai/

A Smart City – according to the new European Smart City 4.0 model [17, 18] can be defined:
"as a city well performing in 6 key fields of urban development, built on the 'smart' combination of endowments and activities of self-decisive, independent and aware citizens"

The European Smart City 4.0 initiative proposed different Smart Cities to be compare on 6 different indicators as can be seen with an sample result for 3 Smart Cities – Bochum, Germany, Aarhus, Denmark and Gdansk, Polen in Figure 10.4.

Essentially "people" are covered by one parameter – and this param-eter related to people becoming smart via technology. However Smart in this context measured as on Education, Lifelong learning, Ethnic plurality, Open-mindedness.

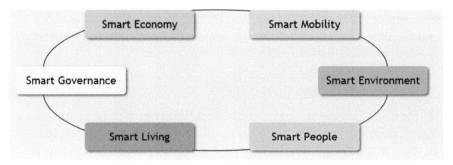

**Figure 10.3** Key fields of Urban Development. *Source:* europeansmartcities 4.0 (2015) [17, 18].

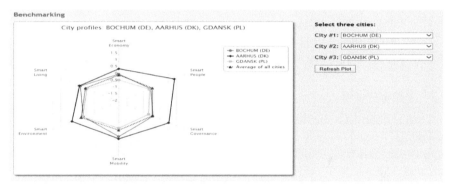

**Figure 10.4** Comparement of 3 Bochum, Aarhus, Gdansk Smart cities on 6 key parameters of Smart city development. *Source:* European Smart Cities 4.0 (2015) [17, 18].

Another approach and viewpoint to Smart City is about the integrated system of different cyber physical systems and various innovative technologies such as ICT, IoT, ITS, Smart Grid, Cyber Security, etc.

## 10.3.9 Information and Communication Technology (ICT)

ICT acts here again as one of the prime enabler for making cities smart. Well planning and execution of ICT infrastructure plays a vital role in the development of both Greenfield and Brownfield smart cities. ICT planning is claimed to help in using available resources in intelligent and efficient way resulting in cost and energy savings with reduced environmental footprints. For the successful and sustainable implantation of ICT for the development of smart city, the essential services include deployment of massive broadband networks, appropriate usage of smart devices and agents, developing urban smart spaces, developing web based applications and e-services, and opening up Government data initiatives under right to information [55].

### 10.3.10 Internet of Things (IoT)

For transparent and seamless interconnection and integration of large number of heterogeneous cyber physical systems and providing them various data subsets through different open data access services, IoT is here claimed to be the best technology. Urban IoTs support smart city development by using advanced communication technologies for supporting and providing value added services to the citizens so that they can actively participate and contribute to good governance. IoT can help in many tasks such as monitoring structural health of ICT infrastructure, city waste management, air quality monitoring, noise monitoring, traffic congestion, city energy consumption, smart parking, smart lighting, automation and cleanliness of social spaces [4].

### 10.3.11 Smart Grid

Smart Grid is claimed to be the most effective long term approach in meeting city's energy demands with cost efficient solutions. For sustainable smart cities, it is very much essential to provide assured electricity supply to every citizen and various ICT services with less environmental footprints. Use of renewable energy sources such as photovoltaic cells, wind power, tidal power, and vibration energy is claimed to make electricity grid smarter and Smart Water Meters are giving information about usage, quality, leakage identifications and preventive maintenance. For energy management, smart electricity meters helps monitoring usage, energy efficiency improvement and reduction in losses and delays in fault detection and diagnosis in the distribution lines [16].

### 10.3.12 Intelligent Transportation System (ITS)

ITS are claimed to enable rapid transit system, shady streets for pedestrians, Micro metropolitan to semi individual use, Electric Vehicles, City Cargo for CO2 emission reduction, clean public transport, cycle paths, canals. ITS demands walkable city planning, cycles, trams, trains, city buses powered from bio-fuels, trains powered by electricity generated from renewable energy sources, car use is reduced, Ethanol is used as a fuel for vehicles. ITS proposes to create corridors for automated vehicles to move city residents, electrify city fleets, and equip vehicles with connected vehicle technologies [16].

### 10.3.13 Cyber Security

The Guardian [58] reported in 2014 that – "The truth about smart cities will be that in the end, they will destroy democracy". The Guardian claimed that "The increased interfaces among various cyber physical systems in the smart city paradigm globally interconnected physical, economic, social and political sub systems will increase the susceptibility of a smart city's cyber security". The Newspaper argued that "due to the continuous big data transactions, cyber threats would become multiplied. Security provision is a continuous process, which gets modified according to previous attack experiences". The latest cyber attach in May 2017 showed how vulnerable large ICT, IOT systems – hereunder Smart Cities healthcare and energy systems can be [78] – some of the major core components in Smart Cities [10, 13, 17, 18, 42].

## 10.4 Business Model EcoSystem (BMES) Related to Smart Cities

Taking all the above mentioned definitions and considerations – both positive and negative "cost" and "values" – of Smart Cities, it could now be valuable to discuss the Smart City Approach in the context of BMES and what is essential

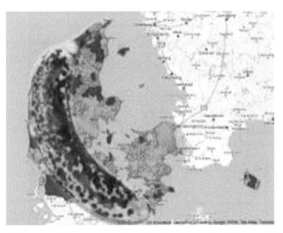

**Figure 10.5** The rotten banana – rural and village areas in Denmark. *Source:* [33].

to keep a Smart City sustainable in a long term perspective business and BMES wise.

Today there is increased awareness of BMs [40, 46, 28, 8, 69, 56, 7, 47, 33] However, with increased use and research of BM the fuzziness on how the BM really is constructed and defined has increased even more. Everybody seems to have their language and terminology – and academia's still lack's to agree on a common language and terminology [57]. From acknowledged academic works with the dimensions of a Business Model and business, we found some generic dimensions that support the idea that any BMES could be defined by 7 generic dimensions and we believe that a Smart City could be compared to a BMES – we propose it called a Smart City BMES (SCBMES).

### 10.4.1 Value Proposition Dimension of a SCBMES

All BMES – and thereby as we propose in this paper – also all Smart Cities – offers values to either BM's inside the BMES and/or to BM's outside the BMES. However most focus – in all the cases we studied – is SCBMES value propositions offer to BM's inside the BMES – only very few Smart City concept focus on the value proposition to BMES outside the BMES.

The BMES value proposition seems to be a "mirror" of the BM'S value propositions individually and together inside the BMES. We define these as the BMES value proposition offered to other BM's either offered as one BM's to another or more BM's together as a shared value proposition of the BMES. Value propositions from a BMES can be offered in the form of products, service and/or process of services and products – and in these cases physical, digital and virtually. Value propositions offered by SCBMES are therefore – we propose – the sum of different BM's physical, digital and virtual value propositions. This increase potentially tremendously the value propositions of Smart Cities – however not to other BMES – if they are not really offered and made available to other BMES by the SCBMES. The "The Smart City" becomes thereby and therefore an "island" – and an isolated BMES.

### 10.4.2 Customers and/or User Dimension of a BMES

BMES – and thereby SCBMES – serves customers or/and users.

> *"A successful* BMES – and thereby a SCBMES – *is one that has found a way to create, capture, deliver, receive and consume value for and from its users and customers – that has found* "a way" *to*

*"help" users and customers to get an "important job" done – "solve pains" and "create gains" for its "users" and "customers". "It's not possible to invent or reinvent a BMES without first identifying a clear customer and/or user base".*

Here, we draw a distinction between customers and users to a SCBMES – and in this case a Smart City. Customers to the BMES pay with money – *"there is no BMES marked – Business of a BMES – if there is no customers and they do not pay"* (Kotler 1983), whereas users to a BMES pay with other values [66] than money. Business Model theory has mainly considered the business model related to customers. Our investigation finds that Smart City theory and Smart City approaches have mainly also just focused on the users – forgetting about the customers. However, as we have verified in our previous research [37] users can be and are highly valuable to BMES by "paying" with other values (Facebook, Google as good excamples here).

### 10.4.3 Value Chain Functions [Internal Part] Dimension

Any operating BMES – also SCBMES – has functions that it has to carry out and which enables the SCBMES to "offer" the value propositions to its users and customers. A value chain function list includes primary and secondary functions of a BMES. Primary functions can be – e.g. inbound logistics, operation, out bound logistics, marketing and sales, service – and secondary functions – support functions – e.g. procurement, human resource management, administration and finance infrastructure. BMES however also have to focus on what we call a tertiary function – *business model ecosystem innovation function.* Elsewise we claim that the BMES and the Smart City will potentially vanish or become a "SmartCity" without life, innovation, learning and sustainability. Anybody, any country or any city can establish a Smart City and "call" themselves a Smart City. Few – however – will be able to make the Smart City be successful – without continues improvement, continues business model innovation and learning. This does not mean that the SCBMES has to carry out all value chain functions by itself – but – as we propose – could with advantage have some functions carried out by network partners and/or even by BMES or businesses outside the smart city. This would result in bringing in new valuable values and competences to the BMES.

So – any operating BMES – also SCBMES – needs to have someone to carry out these above mentioned value chain functions to enable a SCBMES to create, capture, deliver, receive and consume value proposition to and from its users, customers, competences and network partners. Either these can be carried out by its own competence or users, customers, and network partners even other network partners of other SCBMES.

### 10.4.4 Competences Dimension

In BM's we have earlier [37] inspired by Prahalad and Hammel [49] divided competences in to four groups – technology, human resource, organizational system and culture. In a BMES we consider also the 4 competence dimension to be technology, human resource, organizational system and culture with the different BM's "pooling" their competences. In our research, we found that different Smart Cities have different technologies, different human resources, different organizational systems, different culture – and they are on different levels as shown in the European Smart City index [17]. The pool of these competences forms the "shared competences" available in the BMES – in this case the Smart City. The Competences – we claim – are one of the raw materials for future BMI of the SCBMES.

### 10.4.5 Network Dimension

We acknowledge that some BMES – and in this case SCBMES – sometimes regard themselves as large enough – The Metropol Smart cities – isolate to other BMES or do not relate, need and want to related to other BMES. We argue that any BMES – also SCBMES – either they want it or not are in a network of BMES – and these networks of BMES's can either be physical, digital or/and virtual [12, 9, 67, 65, 34]. We found that most "successful BMES" is those that has found a way to create value with and for its network of BMES – that has found "a way" to help other network of BMES get "an important job done". In the Smart City area we found many examples of network of Smart Cities:

**European Smart Cities** – http://www.smart-cities.eu/ – which divides mainly the Smart Cities up according to the numbers of inhabitants according to the Urban Audit Database [17, 18], **Indian Smart Cities** – [21, 42], **China Smart Cities** [10]. Many SCBMES as we found in our research do not really understand and often do not acknowledge value which they receive from other BMES before it is too late and they are in risk vanishing.

### 10.4.6 Relation Dimension

Business models are related through tangible and intangible relations [50–52, 3, 5] to other business models [27, 3, 53, 37]. Businesses are related through strong and weak ties [23]. As BMES – and hereunder SCBMES are ecosystems and therefore construction of BM's it seems also obvious to argue that these SCBMES are also related through tangible and intangible relations – and also with strong and weak ties. BMES send value propositions to other BMES through relations and receive value propositions from other BMES through relations. Relations can be one to one or one to many. Relations can be visible and invisible to humans or machines [35]. Tangible and intangible relations are used in the BMES to deliver and receive values [5]. BMES relate their BM's value proposition, users/customers, value chain functions, competences and network through relations. Relations are used for creating, capturing, delivering, receiving and consuming values. It is obvious also – that if BMES are not connected – "related" to other BMES or BM's then value transfer is impossible.

### 10.4.7 Value Formula Dimension

Any BMES uses some kind of a formula – or actually often many kinds of formulas – to calculate the value it offers to the BMES or other BMES. The value formula is a formula that shows how the value proposition delivered are calculated by the BMES. The results of these calculations is value formula either expressed in money or/and other values. In our research of Smart Cities we found very many value formulas – but also some that had not considered the value formula – which indicates that some Smart Cities are not "fully Baked" businesses. It was very clear that most Smart Cities were calculating on other values than money – which may fit the users.

Several have documented that BMES operates and is influenced by its BMES environment – external environment factors. We address that these external environment factors – political, economic, social, technical, environmental, legal conditions and competitive contexts and environment dimensions are important and critical to any BMES survival and growth – also SCBMES. Therefore, we do not believe and we do not think – it is an illusion – that Cities and Smart Cities in the long run can live as "an island" [27] or isolate in the past with "physical walls" and now projected with "digital walls". Smart Cities will in the long run have to relate to other BMES and the environment outside. Further other BMES and individuals have and always will try "to relate" to the Smart City BMES – in this case and in our days "BMES of cybercrime".

The above mentioned seven dimensions of a SCBMES are equivalent to the overall model we propose to how any business and business model

**Figure 10.6**    Security and attachs on "Smart Cities" in the time of medieval and in the time of cyber and smart cities.

is constructed [37]. The seven dimensions – we propose – should also be considered by any SCBMES. However, there is a difference between the way businesses want to run their operations in a BMES – in a SCBMES – and the seven visionary dimensions of a business and Smart Cities. We found that Smart Cities and their related businesses run their BM's differently in Smart Cities – even different in different Smart Cities – and most businesses have more than one BM in a Smart City. In other words, the businesses they described via the seven dimensions are different to how they actually run their business models in the Smart Cities. Some of these business models are close to their original description of the seven dimensions but others were often very different to these. This often challenges the survival and growth of a SCBMES – but it also drives the development, technology, humans, organizational system, culture and vitality of a SCBMES – in this case a Smart City. If more Business begin to run their BM's not in "sink" with the Smart Cities overall vision, mission and goals of the 7 dimensions then the Smart City can be challenge and eventually be disrupted, torn apart and vanish.

This places our attention to the "download", "see" and "sense" approach to SCBMES in the perspective that those SCBMES that we studied have more BM's that are different. Any SCBMES have to be aware of these and their development. We address the importance of continuously investigation of the SCBMES and their BM's and not least their BMI. Any BMES has to have a clear "picture" of the distinction between the "visionary model" of the SCBMES – in this case the Smart City and the individual BM's of business that are actually operating in the SCBMES ("AS IS" BM) and are intended to be operated ("TO BE" BM) in the SCBMES. Herein we believe lays the "seed" to SCBMES longterm survival as if they are not "in sink" with the manifold of BM's then the SCBMES eventually is in risk of falling apart and maybe vanish. Further if all SCBMES BM dimensions are not clear and defined continuously then the SCBMES is in risk of falling apart, competed or even be disrupted by other BMES that have a clearer or better BM.

This observation together with inspiration from Abell's and Hamel' original definitions and framework of "The core Business" [1], "The core competence" [24] made us draw an analogy to the definition of "the BMES" and Smart Cities – as the BMES context – and visionary level states of BMES seems to be related to the seven above mentioned dimensions. This means that we propose that one answer to the question – What is a Smart City? could be that the core of a Smart City BMES (SCBMES) refers in this perspective to:

> "How a Smart City BMES are constructed and intends to operate its "main" and "essential" business related to the seven BMES dimensions – value proposition, user and/or customer, value chain [internal functions], competence, network, relations and value formula."

We believe that any SCBMES can be defined related to the 7 dimensions but we also argue that a SCBMES different BM's cannot be explained by just one BM's – "the core business model" of the SCBMES – but would with preference be better to be explained by different BM's in the SCBMES – however, still each with seven dimensions, but with different characteristics – components. In our research, we found many examples of different BM's operating in a SCBMES, which indicates the existence of more BMs in a Smart City BMES.

As a consequence, we propose that any SCBMES can be said to have more BMs offered by different businesses – the multi-business model approach [34] – which are more, less or not aligned with "the core business model" of the SCBMES. However, any of these BMs can be defined as related to an overall generic SCBMES BM consisting of seven generic dimensions. Each of the seven dimensions of a SCBMES addresses some core questions in relation to each individual SCBMES's dimensions characteristics and logic.

## 10.5 Worldwide Smart City Use Cases Combined with the SCBMES Approach

All over the world, the smart city movement is taking place. Table 10.1 enlists some of the use cases of Smart Cities in the world we studied with some of their core vision, goals, strategies and SCBMES dimensions compared to the BMES approach [38].

**Table 10.1**   BMES's dimensions characteristics and logic

| Core Dimensions in a Smart City BMES – (SCBMES) | Core Questions Related to Dimensions in a Smart City BMES – (SCBMES) |
| --- | --- |
| Value proposition/s (products, services and processes) that the Smart City BMES offers (Physical, Digital, Virtual) | What value propositions do the Smart City BMES provide? |
| Customer/s and Users that the Smart City BMES serves – geographies, physical, digital, virtual). | Who do the Smart City BMES serve? |
| Value chain functions [internal].(physical, digital, virtual) that the Smart City BMES carries out | What value chain functions do the Smart City BMES provide? |
| Competences (technologies, HR, organizational system, culture)(Physical, digital, Virtual) that the Smart City BMES have | What are the Smart Cities BMES competences? |
| Network – Network and Network partners (strategic partners, suppliers and others (Physical, digital, virtual) that the Smart City BMES includes or collaborate with in its BMES | What are the Smart Cities BMES networks? |
| Relations(s) e.g. physical, digital and virtual relations that the Smart City BMES have. | What are the Smart Cities BMES relations? |
| Value formula (Profit formulae and other value formulae. (physical, digital, virtual) that the Smart Business BMES use. | What are the Smart Cities BMES value formulae? |

These SCBMES approaches we can now enlists also to where they are **focusing most** related to BMES dimensions compared to the internal and external SCBMES dimensions and approach [38].

## 10.6  Reflection and Discussion

The paper commence the journey of building up a "language" on Smart City BMES (SCBMES) on behalf of case studies within the American, European, Indian, China, Korean and South America Smart City communities and SCBMES. As the main case study of the 10 elected SCBMES projects shows, there is definitely not one way to create and run a SCBMES. It is also very clear that the SCBMES are "seeing", "sensing", "thinking", "acting" and "doing" differently – mostly focusing on their own SCBMES and not particularly considering External BMES and their environment outside the SCBMES. It means that the SCBMES are developing for their own advantage and benefits – mostly from a user perspective – and not particularly thinking about who is or should be the customers of the SCBMES. This becomes very clear when analyzing the value formula, as most SCBMES are not focused on monetary value formulas – but on other value formulas. Futher they are not very clear about their value formula calculation. Most SCBMES seems to be like an innovation project in a business that are without clear KPI's. This indicates that SCBMES are not thought generally from a profit oriented business point of view – except as far as our investigation shows us – SCBMES Shanghai. Further if seems as if most SCBMES are focusing on primary functions and very few on tertiary functions (BMI). There is an overweight on digital value propositions – mostly services, together with digital network and relations – meaning that the physical network and relations – and the humans are in many SCBMES not particularly in focus.

The value proposition of the SCBMES are mostly focused on service and digital service – but in general physical and virtual products is not thought about and not integrated with the service value proposition. This means a very large potential of SCBMES seems not to be used yet – and that user and customers that cannot access digital the SCBMES are left behind.

Opposite to previous Smart City definitions we propose a new terminology – where the SCBMES is defined as related to both "AS IS" and "TO BE" business BM's. We propose that any SCBMES as define to 7 dimensions (value proposition, user and customers, value chain function, competence, network, relation and value formula. We propose that SCBMES have more focus in future SCBMES innovation and development on an integrated SCBMES – with integrated physical, digital and virtual layers of the SCBMES. We propose that the SCBMES should focus more on the business side and BMI part of the SCBMES – opposite to previous terms using more narrow focus on ICT, IOT, Security, intelligence, health, mobility and energy. Technology and Business have to work much more together in symbiosis – also in the SCBMES – otherwise the Smart City concept will face some mayor risks – not becoming sustainable in the long run – and definitely not business wise.

SCBMES must also bring the people and the humans – users, customers, employees and networkpartners more into the SCBMES. Many Smart City projects seems to forget the human – the people – and focus more on "the Smart ICT" and all the data available – forgetting about real wants, needs and demands of the humans in this context.

**Table 10.2**  Some Smart City use cases in the world related to BMES Approach

| S. No | Smart City | BMES Vision, Goals and Strategy | BMES – Value Proposition, Users and Customers | BMES – Network | BMES Value Formula |
|---|---|---|---|---|---|
| 1 | Masdar Abu Dhabi Arab Emirates https://www.youtube.com/watch?v=FyghLnbp20U | Be both a Physical and Digital BMES<br><br>To become a **Green Smart City** via emerging and advanced global clean-technology.<br><br>Aims to be one of the world's most sustainable Smart Cities powered by renewable energy.<br><br>Innovation in sustainable products, services, processes and business models | Electricity supplied by a concentrated solar power plant will power a transit system that is 100% electric, as well as seawater desalination plants.<br><br>Each household will be connected to a network for monitoring energy consumption | Industrial networks are designed to mimic the cyclical behavior of natural ecosystems. | State and private funded – not clear yet how the value formulae for the Smart City BMES will be constructed in the future |
| 2 | Amsterdam Netherlands, North Europe | Physical and Digital BMES<br><br>Aim at developing a **greener and healthier city** to live in. | Air quality in the city will be improved through continued efforts to encourage the use of electric transport, by stimulating smart distribution processes and extending low | In Smart City knowledge Network with Copenhagen, Barcelona | EU, State, Municipality and to some extent private funded – not clear yet how the value formulae for the Smart City BMES will be constructed in the future |

|  |  |  |  |  |
|---|---|---|---|---|
| | | Aim at achieving **energy saving** and accelerating the process of connecting existing houses to district heating systems.<br><br>Construction market "pushed" to Build greener properties | emission zones to include more types of vehicles.<br><br>Schools that are greener and healthier places to work | |
| 3 | Stockholm Sweden<br><br>http://international. stockholm.se/govern ance/smart-and-connected-city/ | Physical and digital BMES<br><br>A strategy to further develop the smart city through coordination of the City's work on **digitalization**.<br><br>The strategy has been developed together with the Stockholmers – the citizens<br><br>Make Stockholm a smart and connected city, via innovation, openness and connectivity | A city that utilizes digitalization and new technology to simplify and improve the life for its residents, its visitors and businesses.<br><br>To offer the Stockholmers the highest **quality of life** and the **best environment for business**<br><br>In the smart city, new **smart services** are constantly created to make the city even better. | In Knowledge Network with European Smart Cities<br><br>Public and private network partners | EU, State, Municipality and to some extent private funded – not clear yet how the value formulae for the Smart City BMES will be constructed in the future |

(*Continued*)

**Table 10.2**    Continued

| S. No | Smart City | BMES Vision, Goals and Strategy | BMES – Value Proposition, Users and Customers | BMES – Network | BMES Value Formula |
|---|---|---|---|---|---|
| | | Make the city more economically, **ecologically, democratically and socially sustainable.** | Connectivity, publicly accessible data, IT platforms that can communicate with each other, sensors and other technologies | | |
| | | To take advantage of the opportunities presented by digitalization | | | |
| | | A smart city is a sustainable city | | | |
| 4 | Freiburg Switzerland http://www.smartcity-schweiz.ch/de/intere ssengemeinschaft/ workshops-ig-smart-city/p/4/ | Physical and Digital **BMES** Aim to be a **Green City** Smart Aim at **integrating urban energy and transport systems** Open Data & Smart City | Smart tools and services for integrated urban energy and transport systems Smart data, big data Smart governance and smart citizens | In network with European Smart Cities and other Smart cities in Schweiz | State, Municipality and private funded – not clear yet how the value formulae for the Smart City **BMES** will be constructed in the future |

| 5 | Songdo South Korea https://publications. iadb.org/handle/11 319/7721#sthash.Y Gb3xdrK.dpuf | Physical and Digital BMES Focus on six sectors including **transport, security, disaster, environment, and citizen interaction** while other **services** related to home, business, education, health and car are also being developed. | Specialized service in Songdo includes smart bike services, criminal vehicle tracking and monitoring unusual activities through motion detecting technology Citizen and businesses of Songdo | In network with other global Smart Cities Private-public partnership (PPP) formation | State, Municipality and private funded – not clear yet how the value formulae for the Smart City BMES will be constructed in the future |
| 6 | Shanghai China http://smartcitynews. global/why-shanghai- is-about-to-become-a- truly-smart-city/ | Physical, Digital and Virtual BMES It embraces the strategy of **"Internet plus" – leveraging the interplay between the Internet, manufacturing and commerce** **A smart local ecosystem** | Smart transport, smart grids and the Internet of Things Education services are proliferating through a dizzying range of websites and apps, supporting learning for all ages, as well as connecting teachers, administrators and parents. Learning about almost Anything, anywhere. | Mainly Networks within Shanghai | City and private paid City, Citizen and Business paid |

*(Continued)*

**Table 10.2**   Continued

| S. No | Smart City | BMES Vision, Goals and Strategy | BMES – Value Proposition, Users and Customers | BMES – Network | BMES Value Formula |
|---|---|---|---|---|---|
| | | **A center of technological innovation** that reaps rewards in terms of both **business** results and **cultural transformation** | Artificial intelligence, robots, drones, augmented reality, virtual reality, blockchain, Internet of Things and 3D printing | | |
| | | **Smart and advanced digitalization** | Patient care. Improving electronic medical records (EMR) | | |
| | | **Learning at all ages** | Always on" and entirely at ease with social media, they are adopting new ways of living powered by mobile technology | | |
| | | Improved **health, smart business,** and **mobility** underpinned by **connectivity** | Mobile phone maps, navigation and online ride-hailing services | | |
| 7 | Copenhagen Denmark | Physical and Digital BMES | Citizens of Copenhagen | The cities of Copenhagen, Antwerp and Helsinki have launched an | EU, State, Municipality and to some extent private funded – not clear yet how the value formulae for the Smart |
| | | **Open Data** about Copenhagen, free to use through a portal. | UNDERBROEN is a space for exploration of urban production | | |

| | | | | | |
|---|---|---|---|---|---|
| http://cc.cphsolutions lab.dk/shortcodes.html http://cphsolutions lab.dk/ | *Street Lab* is Copenhagen's testarea for **smart city solutions in real urban space** | and a meeting place for Innovators and disruptors | innovation challenge to co-create the next generation IoT platform for Smart cities. | City BMES will be constructed in the future | |
| | | *EnergyBlock* is the test site of Copenhagen Solutions Labs for Decentralised Energy and Blockchain solutions. | | | |
| | | *The City Data Exchange* will make it possible to purchase, sell and share a broad range of data types between | | | |
| 8 Vienna Austria https://smartcity. wien.gv.at/site/ | Physical. Digital and virtual BMES | Citizens and business of Vienna | In knowledge network with European smart cities | EU, State, Municipality and to some extent private funded – not clear yet how the value formulae for the Smart City BMES will be constructed in the future | |
| | "Providing **energy in new ways** – for instance with **citizens' power plants** | Water management and smart grid infrastructure Waste management and green spaces are all key areas, coupled with innovative energy-efficiency programs. Reducing resource consumption | Public and private network partners | | |
| | **Quality of life: Social housing.** | | Involvement of the internal city bodies and forces, as well as overarching cooperation | Energy-efficient and sustainable economy in the city | |
| | Sreate a supportive and structured Smart City framework | | | | |

(Continued)

**Table 10.2** Continued

| S. No | Smart City | BMES Vision, Goals and Strategy | BMES – Value Proposition, Users and Customers | BMES – Network | BMES Value Formula |
|---|---|---|---|---|---|
| | | Addresses **a cross-section of the entire city** and affects **virtually all areas of** responsibility <br>• **radical protection of resources** <br>• **holistic perspectives** <br>• **a high, socially fair quality of life** <br>• **productive use of innovations/new technology** | "Intelligent traffic solutions, green buildings <br>Smart technologies, systems and concepts | with the associated businesses in the city. | |
| 9 | Rio Brazil <br> <br>http://www.bbc.com/news/technology-22546490 <br>https://publications.iadb.org/handle/11319/7727 | Physical and Digital BMES <br>**Digitalized City** <br>**Smart Transport** <br>**Safety** | Citizen of Rio | In knowledge network with global smart cities in the world <br>Unicef in collaboration with local non-government organization CEDAPS (Centro de Promocao da Saude) | State, Municipality and private funded – not clear yet how the value formulae for the Smart City BMES will be constructed in the future |

| 10 | Mombai India  http://www.smartciti eschallenge.in/files/ dmfile/Draft-Smart-Cities-Proposal-navimumbai1.pdf | Physical and Digital BMES **Smart and eco-friendly habitat.Future destination for smart and sustainable living** Fostering the **Quality of Life of its citizens**, sustainably and responsibly **Physical-Smart, Livable Urban Fabric & Infrastructure** **Systematic investment in infrastructure & technology**. The goal is to increase **sustainability and creating smart infrastructure that promotes livability** To achieve the goal of a **Smart Urban** | Citizens and Businesses of Mumbai Fundamental physical services such as water, energy, Sanitation comforts both citizens & businesses. Attract business investments to The city with its highly efficient and reliable city infrastructure; providing a boost to the economy of the region. | In knowledge network with other smart cities in India | State, Municipality and private funded – not clear yet how the value formulae for the Smart City BMES will be constructed in the future |

**Table 10.3** Focus areas of BMES's dimensions characteristics and logic in 10 Smart Cities of the world

| Core Dimensions in a Smart City BMES | Internal BMES Core Questions Related to Dimensions in a Smart City BMES | External BMES |
|---|---|---|
| **1. Value proposition/s** | | |
| **What value propositions do the SCBMES provide and focus on?** | | |
| **that the Smart City BMES offers (Physical, Digital, Virtual)** | | |
| products | Masdar Abu Dhabi, Amsterdam, Freiburg, Songdo, Shanghai, Vienna, Mombai | None |
| services | Masdar Abu Dhabi, Amsterdam, Stockholm, Freiburg, Songdo, Shanghai, Copenhagen, Vienna, Rio, Mombai | Very few |
| processes | Amsterdam, Stockholm, Songdo, Shanghai, Copenhagen, Vienna, Rio, | None |
| **Customer/s and Users that the Smart City BMES serves – geographies, physical, digital, virtual).** | | |
| **Who do the Smart City BMES serve?** | | |
| Customers | Shanghai, Mombai | None |
| User | Masdar Abu Dhabi, Amsterdam, Stockholm, Freiburg, Songdo, Shanghai, Copenhagen, Vienna, Rio, Mombai | Very few |
| **Value chain functions [internal].(physical, digital, virtual) that the Smart City BMES carries out** | | |
| **What value chain functions do the Smart City BMES provide?** | | |
| Primary Functions | Masdar Abu Dhabi, Amsterdam, Stockholm, Freiburg, Songdo, Shanghai, Copenhagen, Vienna, Rio, Mombai | None |
| Secondary Functions | Shanghai, Vienna | None |
| Tertiary Functions (BMI) | Amsterdam, Shanghai, Copenhagen | None |
| **Competences (Physical, digital, Virtual) that the Smart City BMES have** | | |
| **What are the Smart Cities BMES competences?** | | |
| Technologies | Masdar Abu Dhabi, Amsterdam, Stockholm, Freiburg, Songdo, Shanghai, Copenhagen, Vienna, Rio, Mombai | None |
| HR | Shanghai, Vienna | None |
| Organizational system | Amsterdam, Stockholm, Freiburg, Songdo, Shanghai, Vienna, Rio, Mombai | None |

**Table 10.3**   Continued

| Culture | Amsterdam, Stockholm, Shanghai, Copenhagen, Vienna, Rio, | None |
|---|---|---|

**Network – Network and Network partners (strategic partners, suppliers and others that the Smart City BMES includes or collaborate with in its BMES**
**What are the Smart Cities BMES networks?**

| Physical | Amsterdam, Freiburg, Vienna, Rio, Mombai | Few |
|---|---|---|
| Digital | Masdar Abu Dhabi, Amsterdam, Stockholm, Freiburg, Songdo, Shanghai, Copenhagen, Vienna, Rio, Mombai | (Songdo have some) – the rest very few |
| Virtual | Shanghai, | None |

**Relations(s) – Physical, Digital, virtual relations that the Smart City BMES have.**
**What are the Smart Cities BMES relations?**

| Tangible | Masdar Abu Dhabi, Stockholm, Freiburg, Songdo, Shanghai, Copenhagen, Vienna, Rio, Mombai | Very few |
|---|---|---|
| Intangible | Shanghai, | Very few |

**Value formula – Profit formulae and other value formulae – physical, digital, virtual that the Smart Business BMES use.**
**What are the Smart Cities BMES value formulae?**

| Monetary value | Shanghai, Mombai | None |
|---|---|---|
| Other values | Masdar Abu Dhabi, Amsterdam, Stockholm, Freiburg, Songdo, Shanghai, Copenhagen, Vienna, Rio, Mombai | Very few |

A growing number of smart cities are approaching smart transformation through a series of smaller projects, typically costing a few thousand to a few million Euros. However public sector budgets are often insufficient to take advantage of the benefits of these smart city development and BMI projects. Alternative forms of financing these proejcts from the private sector has to become a priority – or a must – to make Smart City sustainable.

## 10.7 Conclusion

The Smart City BMES (SCBMES) framework was built upon a comprehensive review of academic Smart City terminologies, defintions, cases and business model literature.

The Smart City Business Model Ecosystems (SCBMES) seems to be established and approached very differently from one Smart City to another on the Global Scene. Those Smart City terminologies we know of today seems to be very much focused on one fits all – but the research shows that Smart Cities are very different approached. To "see" and "sense" this may open up for more potentials and even several hidden potentials of the Smart City Concept.

It seems as if the SCBMES around the global world might be able to release much more potential than initial expected and done. This demands however that SCBMES bring "people" and Business into "the concept of Smart Cities" – and do not just see them as "objects". Today many SCBMES have not really brought the people and the businesses into the so-called Smart Cities – and this have left several Smart Cities and Smart City projects in deep challenge – even made them become "Ghost Cities" – which could be argued is not really "Smart" and "intelligent".

Smart City BMES today must change fast related to the context or risk in future SCBMES innovation and development. Otherwise they will expectable vanish and never become sustainable SCBMES and be easy victims of cyber attachs.

SCBMES at different levels and constructions may consider to be established and look very differently to those we seen in the past. A deeper and more holistic approach – all 7 BMES dimensions of the SCBMES should be considered – which will enable us to understand deeper the SCBMES seen in different contexts and could maybe give us some answers to why some Smart City BMES are successful – and others not.

The paper addresses slightly our concern with the difference between "the core business" of the SCBMES – and the variety and strategy of its "AS IS BM's" and "TO BE BM's". If the distance between the core business of the SCBMES becomes too large this can be a reason to why the SCBMES fall apart or are challenge on their survival. Conceptually, the SCBMES is "Smart" and "Intelligent" – but it will be challenge if it do not consider external SCBMES or other BMES and its environment as valuable resources and network partners.

## 10.8 Future Expected Results/Contribution

The study has enlightened a strong demand for testing the SCBMES concept in a larger scale and sample. The next step has been initiated as a bigger quantitative and qualitative empirical-based research to clarify more details of the SCBMES approach and its dimensions. The tests are intended to be a part of several larger EU, Indian, US and China based Smart City research projects

# References

[1] Abell, D. F., "Defining the Business: The Starting Point of Strategic Planning" New Jersey: Prentice-Hall, Inc., 1980.

[2] Adel S. Elmaghraby, Michael M. Losavio, – Cyber Security Challenges in Smart Cities: Safety, Security and Privacy, – Journal of Advanced Research, Vol.5, pp. 491–497, 2014.

[3] Amidon Debra M. 2008 Innovation SuperHighway Amidon, Debra M Elsevier Science ISBN13: 978-008-049-156-1

[4] Assocham Security Report, – Cyber Security: A Necessary Pillar of Smart Cities, India Security Conference, 2016.

[5] Alee Verna and Oliver Schwap 2011 Value networks and the true networks of collaboration Open Source http://www.valuenetworksandcollaboration.com/

[6] Bakici, T., Almirall, E., & Wareham, J. (2013). A Smart City Initiative: The Case of Barcelona. *Journal of the Knowledge Economy, 4*(2), 135–148.

[7] Casadesus-Masanell Ramon and Joan Enric Ricart From Strategy to Business Models and onto Tactics Long Range Planning 43 (2010) 195e215

[8] Chesbrough, H. (2007). Open Business Models How to Thrive in the New Innovation Landscape. Harvard Business School

[9] Child, J & Faulkner D,1998, 'Strategies of Co-operation – Managing Alliances, Networks, and Joint Ventures', Oxford University Press, Oxford.

[10] China Smart City – http://www.smartcityexpo.net/en/ – http://www.smartcityexpo.net/en/ – http://nextshark.com/china-building-first-smart-city-not-beijing-shanghai/

[11] Chourabi, H., Nam, T., Walker, S., Gil-Garcia, J. R., Mellouli, S., Nahon, K., et al. (2012). Understanding Smart Cities: An Integrative Framework. The 45th Hawaii International Conference on System Sciences, (pp. 2289–2297).

[12] Coldmann and Price (1995) Agile competition and virtual Organization Van Nordstrand

[13] Cristina Bueti, Advisor, ITU-T Study Group 5, Overview of the activities of ITU-T Focus Group on Smart Sustainable Cities, 26 March 2015.

[14] Cyberisk, – What Cyber Threats Are Smart Cities Facing? November 2016. http://www.cyberisk.biz/what-cyber-threats-are-smart-cities-facing/

[15] Eleonora Riva Sanseverino, Raffaella Riva Sanseverino, ValentinaVaccaro, Ina Macaione and Enrico Anello, – Smart Cities: Case Studies?, E. Riva Sanseverino et al. (eds.), Smart Cities Atlas, Springer Tracts in Civil Engineering,Springer International Publishing AG 2017.

[16] Escher Group, A Whitepaper for Business, – Five ICT Essentials for Smart Cities, 2015.

[17] europeansmartcities 4.0 (2015)

[18] European Union and Smart Cities – http://ec.europa.eu/eip/smartcities/

[19] Felipe Silva Ferraz1,2, Carlos André GuimarãesFerraz, – Smart City Security Issues: Depicting information security issues in the role of a urban environment?, 7th IEEE/ACM International Conference on Utility and Cloud Computing, 2014.

[20] Fogg B.J. (2003) Persuasive Technology: Using Computers to Change What We Think and Do (Interactive Technologies) 1st Edition

[21] Fri, 2017-01-27 12:20 – SCC India Staff http://india.smartcitiescouncil. com/category-news

[22] Fri, 2017-01-27 10:12 – SCC India Staff http://india.smartcitiescouncil. com/article/scc-niua-and-wd-organize-round-table-discussrole-surveill ance-india

[23] Granovetter, M. S. 1973, "The Strength of Weak Ties", The American Journal of Sociology, vol. 78, no. 6, pp. 1360–1380

[24] Hammel 2001, Value Network

[25] Hernández-Muñoz, J. M., Vercher, J. B., Galache, L. M., Gómez, M. P., & Pettersson, J. (2011) Smart Cities at the Forefront of the Future Internet. In J. Domingue, A. Galis, A. Gavras, T. Zahariadis, D. Lambert, F. Cleary, et al. (Eds.), The Future Internet Assembly 2011: Achievements and Technological Promises (pp. 447–462). Budapest, Hungary.

[26] Help Net Security, – Smart cities face unique and escalating cyber threats?, October 2016. https://www.helpnetsecurity.com/2016/10/20/ smart-cities-cyber-threats/

[27] Håkansson, Håkan & Snehota, I. (1990): No Business is an Island: The Network Concept of Business

[28] Johnson M.W., Christensen, M.C. and Kagermann, H. (2008) Reinventing your business model, Harvard Business Review, vol. 86 No. 12, pp. 50–59

[29] Kimberly Klemm, Industry Technical Writing and Editing, energy-central, – Smart Energy, Smart Grids and Smart Cities in Review?, May 015 http://www.energycentral.com/c/tr/smart-energy-smart-grids-and-smart-cities-review

[30] Kevin Dopart, USDOT, Intelligent Transportation Systems Joint Program Office, – Beyond Traffic: The Smart City Challenge?, Dec 2016. http://www.its.dot.gov/factsheets/smartcity.htm

[31] Komninos, N., Pallot, M., & Schaffers, H. (2013). Special Issue on Smart Cities and the Future Internet in Europe. Journal of the Knowledge Economy, 4(2), 119–134.

[32] Kotler, P 1984 Principles of Marketing Prentice Hall

[33] Krcmar Helmut 2011Business Model Research State of the Art and Research Agenda

[34] Lindgren, P. (2011). NEW global ICT-based business models. Aalborg, Denmark: River.

[35] Lindgren, P. (2012) Towards a Multi Business Model Innovation Model. / Lindgren, Peter; Jørgensen, Rasmus. Journal of Multi Business Model Innovation and Technology 1 edition River Publisher

[36] Lindgren Peter, Morten Karnøe Søndergaard, Mark Nelson, and B. J. Fogg (2013) "PERSUASIVE BUSINESS MODELS" Journal of Multi Business Model Innovation and Technology" River Publisher

[37] Lindgren P. & Horn Rasmussen, O., (2013) The Business Model cube Journal of Multi Business Model Innovation 3. Edition River Publisher

[38] Lindgren Peter (2016) The Business Model Eco – System Journal of Multi Business Model Innovation and Technology Vol: 4 Issue: 2 Published In: May 2016 Article No: 1 Page: 1–50

[39] Lindgren, P. & Wuropulos, K. Wireless Pers Commun (2017). Secure Persuasive Business Models and Business Model Innovation in a World of 5G doi:10.1007/s11277-017-4101-y

[40] Magretta J. (2002) " Why Business Models Matter" Harvard Business Review (80:5)

[41] McLeod Saul 2007, updated 2016 Maslow's Hierarchy of Needs https://www.simplypsychology.org/maslow.html

[42] Ministry of Urban Development, Government of India, Smart City-Mission TransformNation, – Mission Statement and Guidelines, June 2015.

[43] M. J. Prabhu, – Let's Create Smart Villages Before Building Smart Cities, The Wire, Agriculture, September 2016.

[44] Mon, 2017-01-23 12:06 – SCC India Staff http://india.smartcitiescouncil.com/article/mumbai-starts-property-mapping-first-city-enablelidar-project

[45] Brandenburger, Adam, and Nalebuff, Barry (1996). Co-Opetition: A Revolution Mindset That Combines Competition and Cooperation ISBN 0-385-47950-6

[46] Osterwalder, A., Pigneur, Y. and Tucci, L.C. (2004) Clarifying business models: Origins, present, and future of the concept, Communications of AIS, No. 16, pp. 1–25.

[47] Osterwalder 2011 http://alex.aaltoes.com

[48] Persuasive Technology Conference 2016 http://persuasive2016.org/

[49] Prahalad, CK and Hamel, G 'The core competence of the corporation', Harv. Bus. Rev., May-June: 79–91, 1990

[50] Provan, K. G. 1983, "The Federation as an Interorganizational Linkage Network", The Academy of Management Review, vol. 8, no. 1, p. 79.

[51] Provan, K. G., Fish, A., & Sydow, J. 2007, "Interorganizational Networks at the Network Level: A Review of the Empirical Literature on Whole Networks", Journal of Management, vol. 33, no. 3, pp. 479–516

[52] Provan, K. G. & Kenis, P. 2008, "Modes of Network Governance: Structure, Management, and Effectiveness", Journal of Public Administration Research and Theory, vol. 18, no. 2, pp. 229–252.

[53] Russels 2012 Presentation Stanford University 2010 at the EU and US FinES Conference on Emerging Business Models River Publisher 2012 and Russels Martha http://www.youtube.com/watch?v=RrEi-gval78

[54] Schaffers, H., Komninos, N., Pallot, M., Trousse, B., Nilsson, M., & Oliveira, A. (2011). Smart Cities and the Future Internet: Towards Cooperation Frameworks for Open Innovation. In F. I. Promises, J. Domingue, A. Galis, A. Gavras, T. Zahariadis, D. Lambert, F. Cleary, et al. (Eds.), Future Internet Assembly (pp. 431–446).

[55] Scholl, N. J., Barzilai-Nahon, K., Ahn, J. H., Olga, P., & Barbara, R. (2009). E-commerce and egovernment: How do they compare? What can they learn from each other? Proceedings of the 42nd Hawaiian International Conference on System Sciences (HICSS 2009). Koloa, Hawaii

[56] Teece, D.J. 2010. "Business Models, Business Strategy and Innovation", Long Range Planning (43:2 -3), pp. 172 194.

[57] Teece David J. (2011) Business Models, Business Strategy and Innovation Long Range Planning 43/2–3

[58] The Gardian 2008 - https://www.theguardian.com/environment/2008/oct/29/climatechange-endangeredhabitats

[59] The Hindu, – Govt. Announces List of First 20 Smart Cities under Smart Cities Mission, Daily News Internet Desk, Jan 28, 2016.

[60] Times of India, – Smart Cities Project is a mass movement: PM NarendraModi, Jun 25, 2016. Government of India, Ministry of Urban Development, – Cyber Security Model Framework for Smart Cities, May 2016.

[61] TINA Vienna – https://www.metropole.at/cover-story-pimp-city/

[62] Tue, 2017-01-24 13:06 – SCC India Staff http://india.smartcitiescouncil.com/article/scc-roundtable-technology-modernisation-saferand-smarter-city

[63] Vandana Rohokale and Ramjee Prasad, Cyber Security for Intelligent World with
Internet of Things and Machine to Machine Communication?, Journal of Cyber Security, Vol. 4,23–40, 2015.

[64] Vandana Rohokale and Ramjee Prasad,? Cyber Security for SmartGrid – The Backbone of Social Economy?, Journal of Cyber Security, Vol. 5,55–76, 2016.

[65] Vervest, P et al., 2005, Smart Business Networks Springer ISBN 3-540-22840-3

[66] Von Hippel (2005) Democratizing Innovation MIT Press, 01/04/2005Wed, 2017-01-25 14:48 – SCC India Staff http://india.smartcitiescouncil.com/article/temples-ring-high-end-security-systems-sri-kashivishwanath-did-it

[67] Whinston, A.B. Stahl, D.O. and Choi S., The Economics of Electronic Commerce, Macmillan Technical Publishing, Indianapolis, IN, 2003.

[68] Zanella Andrea, Nicola Bui, Angelo Castellani, Lorenzo Vangelista, and Michele Zorzi, – Internet of Things for Smart Cities, IEEE Internet of Things Journal, Vol.1, No.1, February 2014.

[69] Zott, C., Amit, R., & Massa, L. (2010). The business model: Theoretical Roots, Recent Developments, and Future Research. University of Navarra: IESE Business School.

[70] Århus Smart City – http://www.smartaarhus.eu/

[71] Copenhagen Smart City – http://www.copcap.com/set-up-a-business/key-sectors/smart-city.

[72] EU – Smart Cities Investment – https://eu-smartcities.eu/content/europe an-commision-invest-%E2%82%AC200m-smart-cities-next-two-years

[73] EU – Smart City – http://europa.eu/rapid/press-release_MEMO-13-884_en.htm

[74] Smart City Report EU – https://amsterdamsmartcity.com/events/lessen-uit-een-slim-amsterdam

[75] Winden van Willem, Inge Oskam, Daniel van den Buuse, Wieke Schrama, Egbert-Jan van Dijck (2017) – ORGANISING SMART CITY PROJECTS LESSONS FROM AMSTERDAM https://drive.google.com/file/d/0Bz0U6OArm0dId0labDJRRDhHUjA/view

[76] Child, John, Faulkner, David, and Tallman, Stephen B. Cooperative strategy 2005

[77] http://demographicpartitions.org/urbanization-2013/

[78] https://www.theguardian.com/technology/2017/may/14/cyber-attack-escalate-working-week-begins-experts-nhs-europol-warn

## Biographies

**Peter Lindgren Ph.D**, holds a full Professorship in Multi business model and Technology innovation at Aarhus University – Business development and technology innovation and has researched and worked with network based high speed innovation since 2000. He has been  head of Studies for Master in Engineering – Business Development and Technology at Aarhus University from 2014–2016. He is author to several articles and books about business model innovation in networks and Emerging Business Models. He has been researcher at Politechnico di Milano in Italy (2002/03), Stanford University, USA (2010/11), University Tor Vergata, Italy and has in the time period 2007–2010 been the founder and Center Manager of International Center for Innovation www.ici.aau.dk at Aalborg University. He works today as researcher in many different multi business model and technology innovations projects and knowledge networks among others E100 – http://www.entovation.com/kleadmap/, Stanford University project Peace Innovation Lab http://captology.stanford.edu/projects/peace-innovation.html, The Nordic Women in business project – www.womeninbusiness.dk/, The Center for TeleInFrastruktur (CTIF) at Aalborg University www.ctif.aau.dk, EU FP7 project about "multi business model innovation in the clouds" – www.Neffics.eu. He is co-author to several books.  He has an entrepreneurial and interdisciplinary approach to research and has initiated several Danish and International research programmes. He is founder of the MBIT lab and is cofounder of CTIF Global Capsule.

His research interests are multi business model and technology innovation in interdisciplinary networks, multi business model typologies, sensing and persuasive business models.

**Dr. Ramjee Prasad** is a Professor of Future Technologies for Business Ecosystem Innovation (FT4BI) in the Department of Business Development and Technology, Aarhus University, Denmark. He is the Founder President of the CTIF Global Capsule (CGC). He is also the Founder Chairman of the Global ICT Standardisation Forum for India, established in 2009. GISFI has the purpose of increasing of the collaboration between European, Indian, Japanese, North-American and other worldwide standardization activities in the area of Information and Communication Technology (ICT) and related application areas.

He has been honored by the University of Rome "Tor Vergata", Italy as a Distinguished Professor of the Department of Clinical Sciences and Translational Medicine on March 15, 2016. He is Honorary Professor of University of Cape Town, South Africa, and University of KwaZulu-Natal, South Africa.

He has received Ridderkorset af Dannebrogordenen (Knight of the Dannebrog) in 2010 from the Danish Queen for the internationalization of top-class telecommunication research and education.

He has received several international awards such as: IEEE Communications Society Wireless Communications Technical Committee Recognition Award in 2003 for making contribution in the field of "Personal, Wireless and Mobile Systems and Networks", Telenor's Research Award in 2005 for impressive merits, both academic and organizational within the field of wireless and personal communication, 2014 IEEE AESS Outstanding Organizational Leadership Award for: "Organizational Leadership in developing and globalizing the CTIF (Center for TeleInFrastruktur) Research Network", and so on.

He has been Project Coordinator of several EC projects namely, MAGNET, MAGNET Beyond, eWALL and so on.

He has published more than 30 books, 1000 plus journal and conference publications, more than 15 patents, over 100 Ph.D. Graduates and larger number of Masters (over 250). Several of his students are today worldwide telecommunication leaders themselves.

# Index